John O'Reilly

The Anatomy and Physiology of the Placenta

the connection of the nervous centres of animal and organic life

John O'Reilly

The Anatomy and Physiology of the Placenta
the connection of the nervous centres of animal and organic life

ISBN/EAN: 9783337095796

Printed in Europe, USA, Canada, Australia, Japan

Cover: Foto ©berggeist007 / pixelio.de

More available books at **www.hansebooks.com**

THE

ANATOMY AND PHYSIOLOGY

OF THE

PLACENTA.

THE CONNECTION OF THE

NERVOUS CENTRES

OF

Animal and Organic Life.

By JOHN O'REILLY, M. D.,

LICENTIATE AND FELLOW OF THE ROYAL COLLEGE OF SURGEONS, IRELAND ; RESIDENT FELLOW OF THE
NEW YORK ACADEMY OF MEDICINE; MEMBER OF THE MEDICO-CHIRURGICAL COLLEGE OF NEW
YORK ; FORMERLY MEDICAL OFFICER TO THE OLDCASTLE WORK-HOUSE INFIRMARY
AND FEVER HOSPITAL.

NEW YORK :

HALL, CLAYTON & CO., PRINTERS, 46 PINE STREET.

1860.

PREFACE.

No one is more fully sensible of the difficult task of unraveling the mysteries connected with the nervous system than I am. No person is more firmly impressed with the conviction that it requires a man of profundity of thought, depth of penetration, expansive intellect, and enlarged views, to illustrate and demonstrate the mode of action of the nervous system, than I am.

No person is more forcibly convinced of his incompetency to give a subject of such extraordinary importance ample justice, than I am.

What, therefore, has induced me to attempt writing on a subject, which, during a period of twenty years, whilst actively engaged in the practice of medicine and surgery, I never bestowed a thought on?

The reply is simply that, being accidentally called by the Section of Anatomy and Physiology of the Academy of Medicine to attend one of its meetings, with the view of forming a quorum, to enable that body to proceed with a discussion on the Anatomy and Physiology of the Placenta, I complied with the request, and casually advanced views which I subsequently found necessary to vindicate, and which I am now anxious to place before the profession *in extenso*.

INTRODUCTION.

It should be of no consequence by whom the true solution of an abstruse question is found, provided it can be proved to be correct.

Every person who enters on an inquiry relative to a difficult subject, with a view to expound it clearly and satisfactorily, is looked upon with suspicion; his assertions regarded with doubt; and whatever he puts forward as fact is received with hesitation.

The high reputation of an author is generally a passport to the public that the subject he treats of is worthy of attention, and demands consideration.

Conclusions and opinions, arrived at by men after superficial reading and study, are not calculated, in many instances, to elevate the reputation of an author, as they cannot appreciate or comprehend the nature of the matter under discussion.

The opinions of the highest authorities on a scientific subject should not be received as authentic, without standing the ordeal of the most rigid scrutiny as to their truthfulness in all its bearings.

It is particularly requested that the reader of the following pages will negative or affirm every paragraph, after mature consideration, without waiting to read the entire chapter.

The great obstacle to the thorough understanding of the nervous system of animal and organic life, presents itself in the want of human intelligence of a sufficiently high standard to comprehend the agency of immaterialism in the operations of materiality.

The several organs of the body are constructed physically with the greatest precision as to their mechanism and component parts, so as to discharge efficiently the duties of the offices which they were destined to perform in the animal economy.

The eye cannot be surpassed in construction, as an optical instrument; the ear, as an acoustic apparatus; or the larynx, as a musical instrument.

It is to be observed, when the eye is injured, vision is either lost or impaired; that the same remark is true of hearing, when the ear is implicated; that the voice is lowered when the larynx suffers; and so, in like manner, when any part of the organic nervous system is treated with violence, the functions of life are deranged, or cease altogether.

The anatomical organization of a dead man is the very type of the living one. Certain characteristics, which are too well known to require enumeration, distinguish the living from the dead; as the former can be instantly assimilated to the latter, without any trace of disorganization being perceptible; it follows thence, as a sequence, that an immaterial or imperceptible agent must have been expelled.

The union, therefore, of an immaterial agent with materiality, scientifically arranged, as is exemplified in the case of the organic nervous system in man, constitutes life.

The severance of the immaterial agent from the material substance, resulting in death, requires elucidation.

The immaterial agent, known as life, is held in existence by the union of the oxygen of the blood with the structure of the organic nervous system. Whatever, therefore, disturbs this process is followed by fatal consequences: as, for instance, a violent blow on the semilunar, the superior cervical, the cardiac, or the central cerebral ganglia; the shock given to one ganglion is instantly communicated to all the others; thereby the animal machinery suffers concussion, and is unable to attract or combine with the oxygen from the blood: thus the immaterial agent, whose existence depends on the harmonious action of these material agents, (to wit, the organic nervous system and the oxygen,) disappears from its abode.

A dead man and one in a sound sleep closely resemble each other as regards communication with the external world: the operations of the immaterial agent, for which the encephalon was established to provide a habitation for, being totally suspended. The distinction between a dead man and one asleep consists in the operations of the immaterial agent in the organic nervous system being carried on in the latter, whilst in the former they are not.

These facts demonstrate the distinctness of the action and duties of the immaterial agents in the animal and organic nervous system.

A dead man, one recently executed, can be made to appear, to a certain degree, like a living one, by substituting an immaterial agent for the one that has departed—namely, galvanism. It is well known that one-wire of a galvanic battery applied to the lumbar nerves of a frog, whilst the other touches the foot, will cause contraction of the muscles.

It is highly important to understand that the Creator has invariably used material substances when carrying into execution immaterial agencies: thus, for instance, man was made of the slime of the earth; it is evident that without Omnipotent power this could not be accomplished. When Moses struck the rock of Horeb with his rod, water came forth; it is equally clear that an all-powerful, invisible agency was in operation.

The most extraordinary instance of the power of immaterialism being combined with materiality is exemplified in the Incarnation. I cannot avoid giving the text from the first chapter of St. Luke, verse 25: "The Holy Ghost shall come upon thee; and the power of the Most High shall overshadow thee. And therefore the Holy ~~Ghost~~ which shall be born of thee, shall be called the Son of God."

A great many look on the Bible as fabulous, whilst numbers cannot comprehend its sublimity, or construe it to meet their own limited or presumptuous views; in a word, ridicule the idea contained in the above quotation from the Gospel, and thus ignore altogether the power of immaterialism over materiality. There are others who endeavor to account for the phenomena of life by close anatomical investigations of the minute structure of the body, or by chemical agency; who disregard immaterialism, as unworthy of scientific research, and hold it derogatory to their reputation for keen examination and learned exposition.

It is necessary, therefore, to remind the classes of persons alluded to, that the power of an immaterial agent over a material one admits of familiar illustration. A bar of steel or iron, which has been magnetized, either by the application of a loadstone or a current of electricity, will attract another bar of iron and communicate a similar power to it. If two bars of iron are magnetized, the north pole of the one will repel the north pole of the other, whilst the north pole of one will attract the south pole of the other.

Could a clearer demonstration be given of the power of immaterialism over materiality? Does not one bar of iron, made magnetic, impregnate another bar of iron? Is not the analogy of impregnation still further borne out by recollecting that a piece of iron rendered magnetic is capable of attracting to itself other particles or scales of iron in its vicinity for a considerable time—thus enlarging itself precisely as the immaterial agent in the fœtus appropriates to itself the nourishment it requires from the mother?

Electricity is an immaterial agent. It can be generated from matter, as is witnessed in the case of the Galvanic Battery; and can be conducted thousands of miles in a second through wires of various kinds of metal, and the phenomena of electricity rendered manifest by the electric spark.

The current of electricity, when the positive and negative wires of a Galvanic Battery are brought in close contact, will convert gold into fumes, or brilliantly ignite charcoal.

The power of the immaterial agent is here apparent; and it must strike the most superficial thinker that the wires are the mere conductors or instruments—a matter susceptible of proof—shown by cutting off the communication between the wires and the battery.

If the wires connected with the galvanic battery are put into a mixture containing gold, arsenic, copper, iodine, and zinc, the metals will be found reduced to their metalline state by the action of the wires. No person will, I presume, attribute this agency to the wires, but must ascribe it to the immaterial agent they conduct or give abode to.

I trust it does not require further argument to demonstrate the mode of action of the immaterial agent, known as electricity, when in combination with a material substance—such as steel, iron, or copper.

If an inanimate immaterial agent, such as electricity must be admitted to be, is capable of producing such extraordinary phenomena, what must a living immaterial agent be capable of accomplishing?

In the former case, the agent is brought into existence by an electrical machine, or galvanic battery; in the other it is not generated, but held in existence through the instrumentality of the union of the oxygen of the blood with the organic matter in the organic nervous system—the nerves acting as conductors of the vital agent as the wires of electricity. In the case of the galvanic battery, the acid plays a conspicuous part by its action on the zinc in generating the electricity; in like manner, the oxygen discharges a prominent part in its action with the tissue of the organic nervous system, in holding the immaterial agent known as Life.

It is exceedingly difficult to understand, and much more to explain, the manner in which the animal immaterial agent is obtained, as well as rendered active, in the encephalon. The operations of the mind are truly immaterial, and can be transmitted to distant parts by nervous bands acting as conductors, as is exemplified when the heart jumps in the pericardium on the receipt of exciting intelligence, or by an act of volition, as is witnessed when a person wishes to flex or extend his toes—the immaterial agent generated in the brain being thus conducted to the parts specified.

The cerebrum, as well as the cerebellum, is composed of gray-and-white matter, folded up in a peculiar and compact manner. If the brain were fully unfolded, it would be found having a layer of gray substance on the outside and white on the inside.

The structure of the brain contains a large amount of phosphorus combined with oleaginous matter, and is thus rendered susceptible of excitation.

It would seem that the folding up of the brain was done for the purpose of affording a large surface or space for the action of the parts required to be carried into operation in producing or rendering active the immaterial agent.

Arguing from analogy, it would appear, in fact, that the arrangement of the gray and white substances is founded on the same principle that exists when a large quantity of electricity is required to be produced—namely, placing two large sheets of zinc and copper over one another, separated by horse-hair, and then coiled up; thus occupying a small space, and easily brought into action by immersion in an acid mixture. So, in like manner, the gray and white substances are brought into operation through the agency of the phosphorus.

This explanation is an hypothesis, but appears to be

borne out to a certain extent by facts. When a man overworks his brain, phosphorus is found in the urine: showing it has been used to excess in mental operation. The brain requires rest from excitation; hence sleep is called for to suspend its operations, and allow the process of recuperation to take place.

Certain parts of the brain are destined for specific purposes, as can be deduced from the connection of the cerebral nerves. The immaterial agent allied with the sense of smell is in one place—vision in another—hearing in a third—taste in a fourth; and so on with respect of all the other functions of the brain. Each has its own place assigned it.

Thus it is, certain parts of the brain preside over particular functions allied to animal life, just as the organic ganglia preside over the functions of the organs with which they are connected, each having a special duty assigned it.

It is a remarkable fact, that whereas the gray substance is placed on the external surface of the convolutions of the brain, that in the spinal cord it is located internally. Equally worthy of notice is it, that the current of the immaterial agent in the brain is from without inward towards the centre or mesial line, whilst in the spinal cord it is from within outward. In both cases the arrangement is in accordance with the connection of the nerves with the brain and spinal cord.

The conclusion from these circumstances would be evidence that the phosphorus acts on the gray substance, so as to produce the immaterial agent, in a similar way that the oxygen unites with the tissue of the organic nervous system in the sustenance of the immaterial agent known as Life.

It would seem that the white fibrous structure of the brain acts as conductors, as also repositories, of the immaterial agent, after being generated. It requires a certain time to study a subject—the mind, while doing so, is immaterial. The result of the operations of the mind is immaterial, and is deposited in the white substance of the brain, to remain there until called for, in some measure similar to the manner in which electricity is procured in the Leyden jar.

It is a very perplexing matter to conceive how the iris contracts and dilates under the influence of the lenticular ganglion—the heart under the cardiac ganglion—the uterus under the uterine ganglion, or the arteries under the organic nerves, derived from the organic ganglia.

, In truth, it would appear preposterous to some persons even to contemplate that such an extraordinary result should take place through their instrumentality.

But analogy will quickly reflect light on this mysterious subject. For instance, if two solid bars of iron, on being magnetized, repel one another, the positive poles being placed in juxtaposition, will not dilatation be the result? If the south of one is attracted by the north pole of the other, will not contraction be the result? Hence dilatation and contraction are observed alternately taking place between the two bars. The bars contain an immaterial agent, known as electricity; the nerves an immaterial agent, called Life. Hence the phenomena in question can be understood.

When a person sees disgusting food, the stomach rejects its contents. How so? The organic nerves act much in the same way under such circumstances that the positive pole of one magnet does when it repels the corresponding extremity of the other.

When two wires of a galvanic battery touch one another at their extremities, a current of electricity takes place powerful enough to convert gold into fumes or brilliantly ignite charcoal. When food, after the process of mastication, gets into the stomach, the gastric juice is secreted by the stomachic nerves, under the influence of the stomachic glanglion, and is capable of dissolving the hardest substances.

The lacteals, under the influence of the mesenteric ganglion, absorb the nutritious portion of the chyle, just as the wires of a galvanic battery will collect gold or silver in a metallic state in a mixture, where the metals are united with compound bodies.

The phenomena just detailed cannot be attributed to the wires in the one case, nor to the nerves in the other, but to the immaterial agents they conduct. A wire which will convey a current of electricity thousands of miles in a second, is many thousand times smaller in proportion to the distance it has to convey the immaterial agent, than the smallest nerve in the body. When a wire is cut across, the electric current is interrupted; divide a nerve, and the immaterial agent is arrested in its progress. A person wishes to flex his index finger; what takes place? The immaterial agent is generated in the brain, passes through the median nerve to the finger. Sever the nerve at the palm of the hand, and the message is interrupted. Bad news conveyed to the mind is sent to the heart, and sometimes interferes with the good results anxiously expected by a surgeon after a capital operation —as every practical surgeon is made cognizant of by the intermission of the pulse; whilst good news is not only communicated to the heart, but to the fœtus *in utero*, by a continued chain of nervous communication: "For behold, as soon as the voice of thy salutation sounded in my ears, the infant in my womb leaped for joy."—(St. Luke, iv., 44.) Here is a good example of immaterialism acting on materiality. Hence, direct telegraphic communication may be said to exist between mother and child.

I will now conclude by remarking, that the manner in which the arterial blood is converted into venous, is by the absorption or union of the oxygen with the ganglia, in which the capillary arteries terminate and the capillary veins commence.

The retinæ of organic nerves on the external surface of the arteries form ganglia at their extremities, through which the blood passing, loses its oxygen, and consequently becomes venous. It may not be amiss

to remark, that *no organic* nerves can be found in connection with the veins.

Hence the difference between the actions of veins
and arteries.

The Anatomy and Physiology of the Placenta.

The anatomy of the placenta, as well as its physiology, appears not to be thoroughly understood.

In *limine* it may be premised, that the uterus receives its supply of blood from the uterine and ovarian arteries, which it returns by the corresponding veins, and that it is furnished with nerves from the hypogastric, sacral and spermatic nerves.

It is quite certain that the placenta is continually receiving a supply of blood from the maternal circulation, by the uterine arteries, and reciprocally transmitting to the parent source the same quantity of venous blood, by the uterine veins.

It cannot be asserted that the branches of the uterine arteries anastomose with the branches of the umbilical veins, inasmuch as the latter cannot be injected from the former.

It is an ascertained fact, that injection thrown into the hypogastric arteries does not pass into the uterine veins.

The experiments of William and John Hunter are conclusive on these points.

It is an anatomical characteristic of arterial trunks to terminate in capillaries, as well as venous trunks to take their origin from corresponding vessels.

It therefore follows, as a consequence, that the uterine arteries, on passing through the placenta, must follow the prescribed law, and end in capillaries; and that on the same principle the maternal veins or sinuses must also owe their origin to capillary commencements.

The same rule holds good with respect to the hypogastric arteries and umbilical vein, as regards the capillary circulation of these vessels.

It must be readily understood that, unless there was a perfect arrangement of the blood-vessels thus instituted, there would be a complete, or rather unavoidable, commingling of the maternal and fœtal blood; that, in truth, the arterial blood from the mother would mix with the venous blood of the fœtus.

It would not be *philosophical* to presuppose any such *complication*.

It is universally acknowledged that the placenta acts a prominent part in the arterialization of the blood for the fœtus.

The best anatomists of the present time consider the placenta a double organ. They suppose this assumption is proved by the injections made of the maternal vessels, on the one side, and the fœtal on the other.

Assuming the correctness of *this theory*, it must be acknowledged that there must be a *vascular communication* between the respective organs.

Where is such *connection* demonstrated to exist? No person has pointed out the line of *demarcation* between the two bodies.

The idea, that the fœtal blood receives oxygen by the process of endosmosis and exosmosis from maternal arterial blood, *presupposes* that the latter is in a condition to afford the requisite quantity of the vivifying agent, and that the uterine arteries, instead of terminating in *capillaries*, and thus containing necessarily blood of a venous character, present themselves in the shape of small *fountains* all over the *structure* of the *placenta*, separated by membranous partitions from the depots of fœtal blood in the *fœtal tufts;* in plain language, the reservoirs of the umbilical vein.

Taking for granted this ingenious *mechanical arrangement* to be the *correct* one, it may be *asked*, what *becomes* of the *carbon* displaced by the oxygen? The maternal blood *cannot* be *giving* oxygen and *receiving* carbon at the same time. In other words, the *oxygen* from the maternal blood would *unite* with the carbon from the fœtal. *Carbonic acid* would be formed. The *question* now *suggests* itself, what *becomes* of the carbonic acid, as there is no *channel* by which it can make its *escape?* There is no *organization resembling* the *bronchial tubes* or trachea to carry it off.

Another *argument* may be supplied against those who maintain that the fœtal blood becomes oxygenated by being *bathed* in the maternal blood, (here I would say, with due deference, that this is a *very soft* mode of *expression*,) that these authorities can-not *point* out the distinction between the fœtal and maternal blood in the placenta.

Now, what is the appearance of the *placenta?* Its name im-plies its form. Does it not *closely resemble a conglormerate* gland in conformation? Is it not *composed* of an *intricate* net-work, of blood-vessels, connected together? Is it not divided into com-partments by *membranous* septa? Does not a section of the pla-centa present a *lobular* or *granular* aspect? Is not the cut sur-face found *covered* with dots of blood? Are there any large caverns or sinuses to be found in its structure? I believe the answer must be in the negative. Can any *particular* set of *ves-sels* be *traced* to their ultimate distribution? Such has not been *yet done.* I now submit that the placenta is *composed* of lobules, which William and John Hunter describe as cells or interstices; that *these lobules* are placed there for a *specific* purpose; that each *lobule* is *composed* of an uterine artery, uterine vein, hypogastric artery, and umbilical vein; that the *change* in the blood takes place in these *lobules;* that the uterine and hypogastric arteries inosculate in the lobule; that the carbonaceous matter is remov-ed from the blood, and finds its way into the uterine veins, whilst the umbilical capillaries convey the *pure* blood to the fœtus.

The analogy of the anatomy and physiology of the liver and placenta, in my opinion, is very *evident.*

Does not the *liver contain* four sets of vessels—namely, the he-patic arteries, hepatic veins, venæ portæ and biliary ducts? Do not these vessels meet in the *lobule?*

Each lobule, according to Mr. Kernan, is *composed* of a *plexus* of *biliary ducts*, of a *venous plexus*, formed by branches of the por-tal vein, of a branch (intra-lobular) of an hepatic vein, and of minute arteries.

The vena porta and hepatic arteries represent the hypogas-tric and uterine arteries, whilst the hepatic vein represents the umbilical vein. The semblance still holds good with respect to the biliary ducts and uterine veins. The former contain the *excrementitious matter* derived from the portal circulation; whilst the latter *contain* the *impurities* of the maternal and fœtal blood, and in this respect may be said to be, to a certain extent, identical with the biliary ducts.

The doctrine I have now advanced may be deemed hypothetical. But do not certain facts go far to *substantiate* its *truth?* Is not the foetus *developed* in *proportion* to the size of the placenta? If the placenta is diseased, does not the *foetus* become *blighted?* Is not every *organ enlarged* in proportion to the work it has to *perform* in *the animal* economy? Would not a simple inosculation between the mouths of the uterine arteries and umbilical veins be sufficient, if the placenta was not destined to *discharge* a given *function* in regulating the quality of the blood for the foetus?

If the foetal blood is oxygenized by the process of endosmosis and exosmosis, how does it happen that a *foetus, contaminated* by a *syphilitic* taint from the *father*, propagates the *virus* to the *mother?*

Is it not a fair deduction to arrive at the conclusion, that the impurities of the foetal blood must not only be discharged from the foetus itself, but likewise from the mother? Is there any provision made for the reception of any such effete matter in the placenta? If such cannot be found to exist, then how is the deleterious material disposed of? The answer appears to be, by the uterine veins. The supporters of the theory, that the foetal blood is purified by the process of endosmosis and exosmosis, declare that a thin membranous septa separates the maternal arterial blood from the foetal blood in the foetal tufts, and that the oxygen passes from the maternal to the foetal blood. Again I ask, what becomes of the impurities of the foetal blood?

Is there another membranous septum separating the foetal arterial blood from the maternal venous blood?

Does the foetal blood, by the process of endosmosis and exosmosis, give off the carbon and extraneous matter it contains to the venous blood?

I submit there are no direct proofs of any such construction of vessels and membranes.

Great importance is attached to the fact of demonstrating that the uterine arteries pass through the entire substance of the placenta. This truth can be no longer questioned. Prof. Dalton has established its correctness. But is there anything so extraordinary in the course pursued by these arteries? Do they not form a vascular net-work in the internal surface of the placenta? Is not the disposition of the uterine arteries in strict accordance with certain defined principles—as, for instance, the venal arteries? Do not the branches of the latter arteries proceed as far as the basis of the tubular bodies, before they commence their minute ramification? Is not the cortical substance of the kidneys almost entirely composed of terminations of capillaries?

Is it not true that the placenta is composed almost exclusively of blood-vessels? Is not cellular tissue required to connect the component parts together, as is the case in other organs?

Is not the pia mater a thin lamella of cellular tissue, permeated by innumerable arteries? Does it not dip in between the sulci of the cerebrum?

Is not the chorion composed of a structure closely allied in structure to cellular tissue? Is it not highly vascular? Does it not send prolongations into the substance of the placenta?

Is there not a strong resemblance between it and the pia mater? It may be objected that the brain is a non-secreting organ.

Is not phosphorus secreted from the brain? When a man overworks his brain, is not phosphorus found in the urine? In corroboration of my views, I will quote a passage which bears on the subject from the distinguished Mr. Kernan:

"But Glisson's Capsule," observes Mr. Kernan, "is not mere cellular tissue; it is to the liver what the pia mater is to the brain; it is a cellulo-vascular membrane, in which the vessels divide and subdivide to an extreme degree of minuteness; which lines the portal canals, forming sheaths for the larger vessels contained in them; and a web, in which the smaller vessels ramify; which enters the interlobular fissures, and, with the vessels, forms the capsules of the lobules, and which finally enters the lobules, and, with the blood-vessels, expands itself over the secreting biliary ducts. Hence arises a natural division of the capsule into three portions—a vaginal, an interlobular, and a lobular portion." Having now endeavored to show the similarity between the chorion, pia mater and capsule of Glisson, I will direct attention to another point. It may be said, if the uterine veins commence in capillaries, how are the uterine sinuses accounted for? This interrogatory must be answered by asking how the cerebral sinuses are formed. In the former case the sinuses are inclosed between the decidua and the walls of the uterus; in the latter, between the walls of the cranium and the dura mater.

When the placenta is removed from the uterus, there is no appearance of sinuses; the same observation is nearly true of the encephalon. In the one case the capillary veins pierce the decidua to reach the sinuses, in the other they penetrate the dura mater to reach the sinuses. When the decidua are torn from the walls of the uterus, the sinuses are laid open. The same remark is applicable to the cranium when the dura mater is torn off from the bone. [1]

What causes the hæmorrhage when the placenta presents itself over the os uteri? The separation of the decidua from the walls of the uterus, and the consequent destruction of the uterine sinuses.

Professor Simpson, in cases of placental presentation, recommends the removal of the placenta.

How does this operation prevent the hæmorrhage? As soon as the placenta is extracted the capillary supply to the sinuses is cut off, the walls of the uterus firmly contract on the fœtus, the uterine veins are thus firmly compressed, and no hæmorrhage can take place from them. With respect to the arteries, they are necessarily lacerated, and do not pour out their contents on the same principle that arrests or prevents hæmorrhage when a limb is torn from the trunk.

It may be said the fallacy of this explanation is shown, inasmuch as hæmorrhage occurs when the entire contents of the ovum are expelled from the uterus. It is well known that if the uterus is firmly contracted there is no hæmorrhage, and that it must be a flaccid state to admit such a casualty to take place. Now, when this is the case, the uterine veins having no valves, the blood regurgitates and escapes into the cavity of the uterus. The compress in the shape of the fœtus being abstracted, must not an artificial plug in the shape of the hand be introduced into the uterus to supply the deficiency specified?

What other proof is there that the capillaries perforate the decidua to pour their contents into the sinuses? The *bruit de soufflet*. Is not this phenomenon caused by the passage of the blood through a constricted orifice to a wider place? Does not the blood, in flowing from the mouths of the capillaries into the sinuses, afford an example of what the bruit depends on? Does not the blood flow from the small mouths of the capillary veins into the large uterine sinuses? Is it not the received opinion, that the bruit occurs in the walls of the uterus?

I now propose to show that there is a continued nervous connection between the fœtus and mother. Dr. Copeland says that the organic or ganglial nerves are chiefly distributed to the very internal membrane of the blood-vessels, for the purpose of transmitting their vital influence to the blood itself. The experiments of Wilson Philip demonstrated that the nervous power acts a promiment part in the capillary circulation.

I will cite the experiments. "Doctor Philip passed a ligature round all the vessels attached to the heart of a frog, and then cut out the heart; on bringing the toes of one of the hind legs before the microscope, the circulation was found to be vigorous, and continued so for many minutes, at length gradually becoming more languid.

"Doctor Philip's next experiment. The web of one of the hind legs of a frog was brought before the microscope, and whilst Doctor Hastings observed the circulation, which was vigorous, Doctor Philip crushed the brain with a hammer. The vessels of the web instantly lost their power, the circulation ceasing. In a short time the blood began to move, but with less force than natural. The experiment was repeated with the same result. If the brain be not completely crushed, the blow increases the rapidity of the circulation in the web."

These experiments, I think, are satisfactory as to the presence of nerves in the capillaries.

The question now suggests itself, How does it happen that if a woman advanced in pregnancy receives a great shock, that the fœtus dies in utero, or is born with an arrest of develop-

ment, or an idiot? If there is no vascular or nervous connec
tion between the mother and fœtus, the explanation must indeed
be difficult. How can the problem be solved? The solution of
the difficulty, in my judgment, consists in recollecting that the
internal coats of the blood-vessels are lined by the expansion of
the organic nerves, which, with the fœtal vessels, inosculate,
and become continuous in the placental lobule, and thus that
the excitement produced in the nervous system of the mother is
directly communicated to the fœtus. It may be objected that
this *modus operandi* of the nervous system is not capable of being
proved.

However, do not the salivary glands of the epicure actively
discharge their functions at the sight of a favorite dish? Does
not the heart throb, or cease to beat, on the announcement of
good or bad news? Do not the kidneys secrete urine when a
person is terrified? Do not these facts incontrovertibly prove
the nervous power existing between the cerebrum and the
ganglionic nervous system? Did not the child leap in the womb
of Elizabeth on the entrance of Mary?[2] Is not this divine truth
fully borne out by physiological research? Is it not evident,
therefore, that the organic nervous system presides over the
organs of secretion? and may it not, *à priori*, be fairly assumed
that the organic nerves effect the depuration of the blood in the
placental lobule?

I am indebted to the writings of Doctor Copeland as well as
Doctor Paine for my ideas on the action of the nervous system.
I cannot pass by the latter authority without saying he is an
ornament to the medical profession in the United States, and
that he must be acknowledged to be the exponent of the great
truths connected with the nervous system, which others have
promulgated as their own discoveries.

It may be objected, as a fatal objection to my theory of the
anatomy and physiology of the placenta, that the marsupial
animals have no placenta. In order to dispose of this difficulty,
I will quote a passage from Professor Owen in reference to the
matter. In speaking of the Kangaroo, he says: "The fœtus
and its appendages were enveloped in a large chorion puckered
up into numerous folds, some of which were insinuated between

folds of the vascular lining membrane of the uterus; but the greater portion was collected into a wrinkled mass. The entire ovum was removed without any opposition from placental villous adhesion to the uterus; the chorion was extremely thin and lacerated, and upon carefully examining its whole outer surface, no trace of villi or vessels could be perceived; detached portions were then placed on the field of the microscope, but without the slightest evidence of vascularity being discernible. The next membrane, whose nature and limits will be presently described, was seen extending from the umbilicus to the inner surface of the chorion, and was highly vascular. The fœtus was immediately enveloped in a transparent amnion. On turning the chorion away from the fœtus, it was found to adhere to the vascular membrane above mentioned, and to which the umbilical stem suddenly expanded. With a slight effort, however, the two membranes could be separated from each other without laceration, for the extent of an inch; but at this distance from the umbilicus the chorion gave way on every attempt to detach it from the internal vascular membrane, which here was plainly seen to end in well-defined ridges formed by the trunk of a blood-vessel.

" When the whole of the vascular membrane was spread out, its figure appeared to be that of a cone, of which the apex was the umbilical cord, and the base the terminal vessel above mentioned. These vessels could be distinguished diverging from the umbilical cord, and ramifying over it. Two of these trunks contained coagulated blood, and were the immediate continuations of the terminal or marginal vessels. The third was smaller, empty, and evidently the arterial trunk. Besides the extensively membranous ramifications dispensed over the membrane, it differed from the chorion in being of a yellowish tint. The amnion was reflected from the umbilical cord, and formed, as usual, the immediate investment of the fœtus."

I think I am justified in stating that Professor Owen's dissection strengthens rather than weakens my position. It proves, in the first place, that the fœtal vessels end in capillaries; it further demonstrates that a secretion takes place from the capillaries; otherwise how account for the yellowish tint? Exhalation takes place from the capillaries, which gives rise to the coloring matter alluded to. The vascular membrane represents the placenta; the intimate connection of the two membranes points out the connection which subsists between the chorion and placenta; and further shows that the chorion forms the structure in which the blood-vessels form their net-work in the placenta.

But let me inquire, does not the human ovum, in passing through the fallopian tube into the cavity of the uterus, simply enveloped in the amnion and chorion, possess an independent existence, inasmuch as it is not attached in any way, for a certain period, to the uterus?

Does it not follow, therefore, that the ovum of any other class of the mammalia may enjoy the same privilege for a given time? But it may be said that the human ovum can prolong existence only for a little time, and that the shaggy part of the chorion, at a certain point, imbeds itself into the decidua, and thus gives rise to the formation of the placenta, through which it receives sustenance from the parent.

Now, this is all true; but is not the ovum of the kangaroo placed in a similar predicament as the human ovum? The kangaroo fœtus, after leaving the uterus, becomes attached to the nipple of the ovum, in a way not known, I believe, to naturalists, and receives nutriment from this source until it is able to maintain an independent existence. Now, it is clear that neither the human ovum nor the fœtus of the kangaroo can prolong life without attachment to the parent. Therefore the sequence is clear, that the same law governs the human ovum as that of the kangaroo.[3]

NOTES.

[1] This may be said to be only half true, as a lamella of the dura mater separates the sinus from the bone. Now, this is of no consequence, as it is merely intended to show how the blood gets into the sinus by the capillaries.

[2] The following passage from the Old Testament bears closely on the subject under discussion, showing how Jacob caused the cattle to be marked by different colors:

"And it came to pass, whensoever the stronger cattle did conceive, that Jacob laid the rods before the eyes of the cattle in the gutters, that they might conceive among the rods."—Genesis 30: 41.

[3] I am indebted to the "Historical Sketch on the Construction of the Human Placenta," by Francis Adams, LL.D., M.D., since writing the above article, which enables me to give a short epitome of the opinion of ancient and modern authorities on the Anatomy and Physiology of the Placenta.

"Galen maintained there was a vascular connection between the mother and fœtus in utero."

"Fabricius affirms that in the bitch, in the cow, and in the ewe, the vascular connection between the uterus and fœtus can be readily recognized."

"Hoboken, De Graef, Malpighi, Harvey, come to the conclusion, that in all the inferior classes of mammals there is no vascular connection between the mother and fœtus; that the latter is nourished by an alimentary juice which penetrates through the lining membrane of the uterus, and is imbibed through the inverting membrane of the fœtus."

"Doctor Alexander Monroe, primus, says: Were I allowed to illustrate the communication between the mother and her child in the womb by a gross comparison, I would say, that the uterine sinuses are to a fœtus what the intestines are to an adult—the uterine blood poured into the sinuses being analogous to the recent ingestion of food and drink. The liquors sent by the umbilical arteries to be mixed with the uterine blood, resemble the bile, pancreatic juice, and other liquors separated from the blood. The umbilical veins, and those on the surface of the chorion, take up the former part of this composed mass, as the lacteal and mesenteric veins do from the contents of the guts; but the grosser parts of the blood in the sinuses are carried back by the veins of the womb, as the excrescence of the gut are discharged by the anus."

"Haller says: I stick not to the belief that the red blood, as such, is brought from the mother to the fœtus, and transmitted from the fœtus to the mother as it were in a circle."

"John Hunter describes arteries about the size of a crow-quill, passing from the surface of the uterus into the placenta, and terminating there in a very fine spongy substance; and that the veins originating from this same spongy substance pass obliquely through the decidua, and communicate with the proper veins of the uterus."

"Fyfe Compendium Anatomy, 1812. In the placenta are to be observed, on the side next the child, vessels forming the principal part of its substance; on the side next the mother, the ramification of the umbilical branches of the uterine arteries, almost of the size of crow-quills, passing in a transluted manner between the uterus and placenta, and terminating in the latter veins, corresponding with the latter arteries, but flat, and of a good size, running obliquely from the placenta to the uterus, and in the substance of the placenta an appearance which has been supposed by many authors to be common cellular tissue, and easily ruptured by ingestion, but which is considered by late authors as a regular, spongy substance, similar to that in the body of the penis. The placenta receives blood from the uterus, and, according to the opinion of modern anatomists, purifies the blood as the lungs do in the adult for the nourishment of the fœtus. The blood is sent by the arteries of the uterus to the substance of the placenta."

"Doctor Lee (Philosophical Transactions, 1832,) states his observation on the result of the examination of six gravid uteri, and many placentas expelled in natural labor, which seemed to demonstrate that a cellular structure does not exist in the placenta, and that there is no connection between this organ and the uterus by great arteries and veins. He declares, in detaching the placenta carefully from the uterus, there is no vestige of the passage of any great blood-vessel, either artery or vein, through the intervening decidua from the uterus to the placenta; nor has the appearance of the orifice of a vessel being discovered, even with the help of a magnifier, on the uterine surface of the placenta; and further, that no cells are discernible in its structure on the minutest examination. He argues against a vascular connection between the uterus and placenta, from the surface of the latter appearing uniformly smooth, and covered with the deciduous membrane, which could not be the case did any large vessels connect it with the uterus, and from the circumstance that in the majority of cases it separated with the least possible force, and without hæmorrhage."

"Doctor Lee changed his opinion in 1842, and fully agreed with the Hunterian doctrine." "Velpean combats the facts put forward by the Hunters: 1st. In extra-uterine pregnancies such an arrangement is impossible; 2d. The placenta at first does not exist, and even until the third month it consists of agglomerated filaments only, and consequently no sinuses can exist between its lobules; 3d. A regular-formed placenta has been found in connection with a fibrous polypus and hardened portion of the womb; 4th. Velpeau has seen the uterine surface of a recently delivered female hard, leathery, and without orifices."

"Wagner's Physiology, by Dr. R. Willis. The whole placenta, and therefore every individual lobule, consists of two distinct parts—the one a contraction of the chorion and vessels of the embryo, and the other a contraction of the membrana decidua and vessels of the uterus."

Baer's description coincides with Wagner's.

Doctor Knox, (*Medical Gazette*, 1840,) declares there are no cells or cavities in the placenta. He speaks of the decidua being interposed between the placenta and uterus. He further speaks of placental vessels penetrating through this decidua until they reached the surface of the uterus, when they floated in one of the venous sinuses of the uterus.

"Professor Goodsir represents a tuft, or a single point of the placenta, where it comes into immediate contact with the uterus." As neither Doctors Carpenter nor Adams seem to understand Goodsir's description, I pass it over.

Professor Dalton, New York, 1858: "I feel confident, indeed, from the facts which I shall immediately mention, that the blood-vessels of the uterus do really penetrate into the substance of the placenta, as supposed by the Hunters, Dr. Reed, and Professor Goodsir, and that they constitute, with the tufts of the fœtal chorion, an equal part of its mass."

It will be perceived that the theory propounded by me is positively sustained by the opinions of Galen, Fabricius, and Haller; is corroborated by the experiments of the Hunters, Reed, Goodsir, and Prof. Dalton, who have demonstrated beyond a doubt that the uterine vessels perforate the placenta; is fully borne out by the observation of Knox, with respect to the decidua being interposed between the placenta and uterus, as well as the decidua being pierced with vessels; is negatively proved by Doctor Lee, who could discern no vessels on the smooth surface of the intervening decidua between the uterus and placenta. Dr. Lee was led to his very extraordinary conclusions by not being able to detect the small mouths of the capillaries perforating the membrana decidua. His views, although they influenced him to arrive at false deductions, admirably support the explanation I have put forward. I cannot help remarking, that it is strange how a man of Dr. Lee's penetration and sagacity could for a moment suppose that a child, a dozen pounds weight, could be formed in the uterus if it received no sustenance from the mother. His statement that the placenta can be removed without hæmorrhage may be true with respect to the dead subject; but everybody knows, who has attended a case of placental presentation, that it is not well founded with respect to the living, when an abundance of venous blood is discharged on every contraction of the uterus by the laceration of the sinuses. Velpeau's objections being overthrown by Dr. Dalton's able experiments, I pass them over. Wagner entertains the same ideas that I do with respect to the placenta being composed of lobules. Doctor Adams observes that De Graef, Doctor Lee, and others, argue that if the placenta was connected to the uterus by blood-vessels of a considerable size, it would be impossible that the separation of the placenta could ever take place without hæmorrhage. The last remark was still more striking in the case of the inversion of the womb. Now, when the uterus is inverted no hæmorrhage can take place. The vessels being firmly constricted at the cervix uteri, what opposes the reduction when in this condition? The answer has just been given.

The blood discharged in placental presentations is decidedly venous, whilst that in the umbilical vein is truly arterial. Such is Professor Barker's opinion, and other competent authorities.

The Connection of the Nervous Centres of Animal and Organic Life.

It is generally admitted that the secretions from all glandular bodies, as well as the skin and mucous membrane, depend on the presence as well as the agency of the organic nerves. Acknowledging this to be true, it must be confessed that it is just as difficult to trace nervous filaments in their structure as in the maternal or fœtal vessels in the placenta. The deduction to be arrived at from the observations just made, is, that if it were possible to remove every particle of the body with the exception of the substance of the organic nervous system, that the outline of the body would remain perfect, as well as a true delineation of all the viscera, as well as the blood-vessels and muscles.

The retina is the termination or the expansion of the optic nerve. The optic nerve takes its origin from the corpus geniculatum externum and a white fasciculus sent from the nates. The pineal gland rests on the tubercula quadrigemina, and is attached to the inner margins of the optic thalami by two bands of white cerebral matter. In color and appearance it bears a strong resemblance to a ganglion belonging to the organic nervous system. The third nerve arises from the inner border of the crus cerebri. It is distributed to the lavator palpebræ and all the muscles attached to the eyeball, with the exception of the superior oblique and external rectus; it communicates by a small branch with the lenticular ganglion. The ganglion is an exceedingly small body. It sends several filaments to the iris, some of which are lost on the ciliary ligament. I am stating facts, without entering into minute particulars.

The movements of the iris show that the nerves which supply it must come from a source endowed with vital intelligence and instinct. Every person knows that the iris adjusts itself to meet circumstances, and that it vigorously contracts when the retina is threatened with danger by too strong a light being thrown on it. If the lenticular ganglion possesses the faculties just stated, and that it is only connected to the brain by a small branch of the third nerve, I think it may be fairly asked why the pineal gland, (which I look on in the light of the chief ganglion of the organic nervous system,) which communicates with the brain itself, should not have similar characteristics. That the mind's eye must see objects pictured on the retina before the iris discharges its functions, is evident from what occurs in cases of amaurosis resulting from a disorganized condition of the brain, the pupil in such cases continuing dilated. It follows as a consequence, that the nerves proceeding from the lenticular ganglion must act in concert with a sound condition of the brain. As the conservatism of life rests with the ganglionic nervous centres, it is expedient to the well-being of the body that the mind should have free communication with the centre of organic life, so as to be capable of imparting to it, as well as receiving from it, the rules for governing the other nervous centres of organic life.

Ocular demonstration declares that the mind and the ganglion operate consentaneously, as is witnessed in the case of the retina when exposed to light or darkness, and the movements of the iris under such contingencies. The division of the communicating branch of the third nerve demonstrates that the ganglion does not act without corresponding with the brain and central organic ganglion. The cornea, in persons who are obliged to look at very small objects, is rendered convex by the action of the muscles receiving branches from the third nerve, and the axis of vision is properly directed by the action of the muscles specified. How is this matter to be accounted for? The mind becomes conscious through the retina that the latter requires assistance; it is not able to afford the aid required; it communes with the central ganglion, (the pineal gland,) receives power, which it directs to the lenticular ganglion; reflex action takes place through the branch of communication of the third nerve;—the muscles are thus enabled to co-operate in harmony with the mind. It follows as a sequence, that the ganglion must have knowledge of what is passing in the mind; and this is further proved by recollecting that the third nerve is a motor nerve, not guided by intelligence, and that neither the fourth nor sixth nerves, which are distributed to muscles whose action would be to distort the eyes from the proper focus, have any connection with the ganglion.

When a person is overwhelmed with grief tears are freely shed; the nose as well as the eyes become suffused; the mind, as in the former instance, communicates its troubles to the central ganglion, and by reflex action through it to the lenticular ganglion, through the nasal branch of the fifth nerve which is distributed to the pituitary membrane, the integuments at the extremity of the nose, the upper eyelid, the tensor tarsi muscle, the root of the nose, the lachrymal sac, and caruncala lachrymalis; hence the source of the tears is easily explained.

When a person is anxious to hear a sound at a distance, the tympanum is rendered tense by the action of the tensor tympani muscle. Before this can take place, the mind must be made acquainted by the portio mollis of the necessity of further assistance; it communicates with the central ganglion. What is the position of the parts concerned? The tensor tympani is supplied with motor branches from the facial nerve; the mind cannot effect anything through it, but the muscle receives branches from the otic ganglion, which is under the influence of the central ganglion, which inosculate with the branches of the facial nerve, and thus the muscle is capable of acting in unison with the mind. Can it be doubted but that the otic ganglion is coequal in intelligence with the mind in regulating the action of the muscle to meet the requirements of the mind?

That such small bodies as the ganglions alluded to should possess so much instinct, is marvelous in the highest degree. It also follows à fortiori, that all the other ganglia are possessed of the same faculties in guiding the actions of the organs they are established to preside over.

A man that has made a speculation in railroad stock in which he is interested to a large extent, is told that the directors are a band of swindlers; that the concern is a mere bubble. The mind becomes deranged; it sympathizes with the central ganglion, with which it is united. Now, I should remark, the cerebrum is placed in the same position with respect to the ganglion that the motor nerve of the third or sensitive branch of the fifth nerves is connected with the lenticular ganglion; the announcement is forwarded by the par vagum to the pulmonic plexus; hurried respiration succeeds to the cardiac plexus, inducing violent palpitation to the solar plexus, causing the stomach to refuse food to the hepatic plexus, producing a secretion of bile; to the mesenteric plexus, causing diarrhœa; to the renal plexus, inducing a secretion of urine to the hypogastric and sacral plexus, throwing the sphincter vesicæ off its guard, followed by wetting of the pantaloons; and, in case of woman a short time pregnant, abortion, produced by contraction of uterus. From this exposition, the fatal consequences attendant on the division of the par vagum must be manifest, as it actually severs the connecting thread of life. This is proved by recollecting, when a man gets a blow on the semilunar ganglion, that death instantaneously follows. The unity of action between the mind and the ganglia is rendered apparent when a man is about making a large leap; the mind surveys the distance; the will acts through the spinal cord, under the reflex influence of the central ganglion; the motor nerves are summoned to action; the intervertebral ganglia send them filaments to accompany and instruct them how to exercise their power on the muscles, acting on the same principle that governs the lenticular ganglion and third nerve; and thus it is that the muscles act in strict harmony with the mind, propelling the body to the distance contemplated.

The inference from what has been stated is clear, that the spinal nerves are a triple compound, composed of sensitive motor and organic filaments, so that wherever one is found, the other two must be present. The nerves of organic life act consentaneously with the will wherever located, provided there is a nervous connection established between them. For instance, when a man is playing on a musical instrument, he has his mind fixed on the music-book; his fingers move rapidly in harmony with his mind. Now, the nerves distributed to the fingers are branches of the brachial plexus; are capable of giving sensation and motion, as well as organic life, to the parts they are destined to supply, in consequence of being connected with the organic ganglia at the roots of the nerves. The otic and lenticular ganglia again explain the phenomenon.

I presume it must be conceded that physiologists, in experimenting on the hind legs of frogs, have committed a most palpable mistake. They appear to have lost sight of the fact that the spinal nerves are a triple compound. I think the peculiarities of the afferent, efferent, and excito-motor nerves can be readily accounted for when the triple compound nature of the spinal nerves is duly studied. The action of these nerves is attributable to the organic nervous filament they contain.

When a person suffers from irritation in the bowels, caused by worms or indigestible food—here I must observe, that the intestines are exclusively supplied with nerves from the ganglionic system, and that the muscular fibres contract without the agency of motor nerves from the spinal cord. This fact proves that the iris does not require to have the ciliary nerves influenced by a reflected action of the tubercula quadrigemina through the communicating branch of the third. To revert to the patient—he is attacked with convulsions; the irritation is propagated to the brain; from thence to the central ganglion, and by reflex action from it to the brain and spinal cord; and next through the motor roots of the nerves to the intervertebral ganglia, the organic filament communicates with the muscles—the result is alternate relaxation and contraction of the muscles. Now, this is an action of the muscles independent of the will, and cannot be imitated by the most strenuous exertions of the will.

Again, a man gets a lacerated wound in the palm of the hand; the first thing he complains of, if he is about to get tetanus, is that he has caught cold, that his throat is sore, and that he cannot swallow; in some time the muscles about the neck and face become rigidly contracted; next, the muscles of the upper part of the trunk, and finally the muscles of the lower part. This condition of the muscles could not be produced by the motor spinal nerves, inasmuch as the muscles would relax; no man can continue muscular exertion over a given period. What happens can be thus explained: the irritation is conveyed through the organic filament to the cervical ganglia, and then the muscles of the neck, which are furnished with nerves from these ganglia, become contracted; here I need scarcely remark the superior cervical ganglion furnishes the pharynx with nerves. All the ganglia in due time become affected, those connected with the spinal nerves, and similar consequences are the result.

When an individual takes an over-dose of strychnine the muscles become rigidly contracted. Here the poison acts on the organic nerves, and through them on the muscles. It therefore follows, when the muscles are paralyzed, that strychnine excites spasms in the muscles—a fact which is too well known to need comment.

A man gets a dislocation of the shoulder; by throwing him off his guard it is easily reduced; but let him see what you are going to do, although he is most anxious to assist your efforts, the muscles will become forcibly contracted, and baffle your attempts to accomplish your object. Here, too, the mind and the muscles act together, through the influence of the ganglia placed at the roots of the brachial plexus.

When chloroform is administered it first acts on the organic nerves in the bronchial ramifications; secondarily, on the brain, known by the delirium that takes place; thirdly, on the central organic ganglion, and from it, by reflex action, on the brain and spinal cord; and through the nerves of the latter on the intervertebral ganglia, and ultimately the filaments which proceed from the latter to the muscles. That chloroform operates in the manner described is made painfully true, when the heart of man, under its influence, ceases to beat, when all human efforts will prove futile in some instances to restore its vital action—and thus demonstrating, by death, that chloroform is able to destroy an organ endowed with the greatest degree of muscular strength.

When a man gets drunk, the intoxicating liquor makes its first impression on the organic nerves in the coats of the stomach; the par vagum announces its presence to the brain, which becomes exhilarated; and after some time it shares its insidious

sensation with the central organic ganglion, which reflects the
intelligence through the brain and spinal cord to the intervertebral ganglia, and from the latter to the filaments destined to
supply the muscles. This description is literally true; when
witnessing the movements of a drunken man, his lower extremities will be observed not to obey his will, his legs cross one
another, he stumbles from side to side, and, if he has imbibed too
much, the muscles refuse to move, and he falls prostrate to the
earth, perhaps to awaken in another world. Here it is evident
that the faulty movements of the muscles, as well as their loss
of power, is caused by the ganglia at the roots of the lumbar
and sacral plexus, and that death is caused by the destruction
of the ganglionic system; post-mortem examination fails to discover any lesion. My friend, Dr. Finnell, who has made such
an immense number of examinations of the dead bodies of drunk-
ards and others, can corroborate this statement.

A man may die of typhus fever, and on the closest post-mortem examination, no lesion whatever can be found. This matter is now susceptible of explanation; the zymotic poison acts
on the organic nerves of the intestines, conveyed to them by the
impurities contained in the water procured in the immediate
vicinity of privies in large cities, or in the holds of emigrant
ships, as pointed out by a recent writer; or it may be inhaled
into the lungs. When the fever is setting in, it is characterized
by a rigor—the patient grows cold and shivers—the heart beats
feebly—the countenance is sunken and depressed. These indications can leave no doubt as to the implication of the organic
nerves. After some time, the organic system recovers from the
shock; indicated by heat of surface, quick action of the heart,
flushed countenance, and some muscular pains. As the case goes
on, the poison is either worn out of the system, or kills the patient by the destruction of the organic nervous system.

A person may labor under chronic hydrocephalus for years,
and his mental as well as vital faculties continue unimpaired.
This appears very extraordinary, but can be accounted for:—
the cranium becomes enlarged, the brain becomes expanded,
the ventricles filled with serum; the pineal gland, or ganglion,
floats in the fluid. Here, it will be remembered, the body of

the ganglion lies on the tubercula quadrigemina, and is not attached to them; consequently, as the ganglion does not suffer from pressure, the functions of organic life are not interfered with.

A child is attacked with meningitis; all the symptoms of acute inflammation are present—the eyes glisten and the pupils are contracted; showing that the central ganglion is suffering from the irritation. After some time, the child is attacked with convulsions, followed by dilated pupils, partial or complete coma, a quick, feeble, and intermittent pulse. In due time, the patient dies; serum is found in the ventricles. What has taken place? Why, previous to the attack, the cavity of the skull was sufficiently large to accommodate its contents, and no more; therefore the skull being incapable, at such a very short notice, to make further room, the serum presses on the ganglion; hence the phenomena, and death of the patient; the vital principle being compressed out of the ganglion.

Serous apoplexy destroys life in the same way as the case just specified.

When a person suffers from compression of the brain, caused by bone, blood, or matter on the right side of the brain, he may recover, with hemiplegia of the left side. This complication is most assuredly puzzling. How does it occur?

If a man is impelled with force from behind, and strikes his breast against a resisting substance, his breast suffers and not his back. If a cushion is placed on the right arm and then firmly pressed against, whilst the left arm is placed against a resisting body, then it is the left arm that suffers from the pressure, and not the right; precisely on the same principle, when pressure is applied to the right side of the ganglion, the force falls on the left, as it comes in contact with the point of resistance. The left side of the central ganglion is thus rendered unable to communicate with the brain, spinal cord, or intervertebral ganglia at the roots of the motor nerves; hence the hemiplegia.

As some persons will not believe anything without seeing it, I will appeal to direct and conclusive experiments to convince them that the pineal gland is the president of the organic nervous system

Obstetricians caution one another, in performing craniotomy, to be very certain to break up all the brain, and to be very particular to destroy the medulla oblongata, to prevent the possibility of the child screaming after being born—a horrible catastrophe, which has occasionally taken place; thus showing that life is not located in the brain, and that the pineal gland, or ganglion, has escaped injury. Be it remembered, you cannot break up the medulla oblongata without killing the ganglion.

Some children are born with the heads very much elongated; others are delivered by the forceps, similarly circumstanced. How does it happen the children do not die while undergoing this malformation of the head? Firstly—because the bones forming the vault of the skull slide and lap over one another, and the brain, not being a vital organ, suffers no damage. Secondly—the central ganglion is placed in the most secure part, to prevent mischief; the bones forming the base of the skull cannot be compressed without extraordinary force.

When a butcher is about killing an ox, he aims the blow at the central point in the forehead, round which the hair is a little curled—a smart blow here destroys life. Why does it do so? Because the shock is communicated directly to the organic ganglion. An adroit butcher will sometimes dispatch a bullock by plunging a knife between the occipital bone and the atlas, thus severing the medulla oblongata. As death is caused by striking the forehead at the point indicated, as well as by dividing the medulla oblongata, it is clear the vital organ must be placed between those parts, and so it is; the central ganglion is just located in this position.

When the head of a turkey is cut off, it will be observed that life is not extinct either in the head or the body, and that each continue to give manifestations of life for some time; here the central ganglion remains intact, and hence the vitality is not instantly annihilated.

The mouths of persons who have been guillotined have been observed to shut and open after the removal of the head from their bodies. This phenomenon is attributable to the central ganglion not being at once destroyed.

Dogs, pigeons, horses, and rabbits, whose vagi nerves were tied, have lived periods varying from two to five and seven days. The vagi nerves arise from the respiratory tract of Bell, between the olivary and restiform bodies; therefore, as death is not caused by the destruction of the principal nerves which arise from the medulla oblongata for a given time, it is evident the vital principle is not situated in the medulla oblongata.

Again, if the lobes of the brain can be destroyed to any extent without destroying life, it follows, as a consequence, that the vital principle must be placed in particular organs, and such is the fact—namely, in the *organic ganglia.*

I hope I have now given ample proof that the muscular fibres of the muscles are furnished with nerves from the organic nervous system; that the muscles owe their enormous power to their connection with the nerves in question. It may be said no nervous filament can be discovered in the muscular fibre; but the same observation is true of capillary arteries and veins, although no person doubts their presence.

When a drop of prussic acid is placed on the tongue of the subject, as Dr. Copeland remarks, he dies before the poison has time to get into the circulation; post-mortem examination discloses that the blood is fluid and dark colored in the blood-vessels, caused by the death of the organic nerves in the internal coats of the vessels.

If a man is subjected to great violence, and sustains a compound dislocation, he will quickly fall into a state of collapse. The surface of the body will grow cold, the heart will almost cease to beat, the sphincters will be relaxed, and if a surgeon is so imprudent as to perform amputation under such circumstances, it is ten to one but the operation will be attended with fatal consequences, inasmuch as the patient's organic nervous system is unable to bear an additional shock.

Surgeons attribute this condition of the patient to sympathy; but it is more—it is direct and continued communication of the nerves in the part injured with the nerves in other parts. How is the cold on the surface of the body to be accounted for? By the vitality of the organic nerves in the skin being impaired in vitality, and by their being unable to eliminate the oxygen from the capillary arteries;—(this is an hypothesis.)

When a person is seized with Asiatic cholera, on the hottest day in summer the body becomes colder than the surrounding atmosphere, the powers of life quickly give way, and the patient falls a victim to the poisonous influence exercised on the organic nervous system. The state of the patient in cholera goes to prove what I have just stated, that the production of animal heat is a vital action depending on the integral condition of the organic nerves.

As some persons may say that all I have said about the pineal gland being the great organic nervous centre of organic life is mere speculation, I must remark that if the wisest man in the world a century ago was shown a galvanic battery, and told that it was susceptible of generating an immaterial agent that could send a message by a small wire from one extremity of the globe to the other in a second, he would look on the individual giving him such assurance as a man laboring under mental alienation. I cannot help remarking there is an analogy between the pineal gland and the battery; the particles of gritty matter resemble the metallic plates, and the gelatinous matter the acid mixture. The Omnipotent Creator, who showed such infinite wisdom and unity of design in the construction of the organs of sense, did not place the pineal gland in such an important locality for a useless purpose. I strongly conjecture that the ganglion is so constructed as to regulate the *aura vitæ*.

It may be objected, that the gland or ganglion cannot receive and give impressions at the same time. The gland is connected with the brain itself; the nerves proceeding from the brain may be said to be continuations of it. The lenticular ganglion receives impressions from the brain through the branch of the third nerve, and sends communications to the muscles of the eyeball by the same nerve at the same instant. This matter, I think, is now fully elucidated; what is true of the one organ is equally true of the other. I now submit that life is centred in the organic nervous system; that the brain and its appendages are attached for the purpose of affording a seat for the organs of sense, intellect, judgment, volition, sensation, and motion. That the ganglionic system is capable of influencing all the functions of the body, as relates to its preservation and harmonious action, as well as the preservation of the species; that the attributes of the cerebro-spinal system are instituted for man's guidance and connection with the world—that whereas man has the power to control his mind, he has no control over his life.

The will is seated in the brain; this is an admitted fact. The will has no influence over the organs of life. The lenticular ganglion is an organ of life, therefore the brain can have no power over it.

The lenticular ganglion is connected with the brain by a small nerve, and acts in concert with it.

It is the law of the organic ganglia to act in communion with one another.

Therefore, as the lenticular ganglion acts in consequence of being connected with the brain, it follows that there must be an organic ganglion located within the brain. And such is the fact; and placed, too, in the very centre of the brain, and in direct communication with it, called the pineal gland, but more worthy of the title of being styled the president of the organic nervous system.

That there is a close and intimate connection between the organs of animal and organic life, is a matter that cannot be disputed, and such close union is necessary for the well-being of man. Life is the special gift of the Deity.

" And the Lord God formed man of the slime of the earth, and breathed into his face the breath of life; and Man became a living soul."—GENESIS, chap. v., 7.

Having traced the effects of the nervous system to their final cause, I will conclude. I could have entered more fully into details, but I deemed it would be superfluous to do so, having found the master-key to unlock the difficulties connected with the nervous system.

POSTSCRIPT.

Since writing the foregoing article, I have read Bernard's experiments on the nerves entering the submaxillary gland. His remarks, with respect to the blood becoming arterial, and the blood-vessels becoming dilated when the tympano-facial nerve is acting, and consequently the secretion of saliva is susceptible of explanation precisely on the same principles as those governing the connection between the nasal branch of the fifth and its communication with the lenticular ganglion; in the one case, tears are secreted; in the other, saliva. With respect to the blood-vessels contracting and the blood becoming venous, when the tympano-facial nerve is in a state of quiescence, being caused by the action of the organic nerves derived from the carotid plexus, it is clear these nerves preside over the circulation of the blood in the gland; the former condition of the nerves described is destined to make the mind act in concert with the vital action of the organic nerves, whilst the latter is instituted for the preservation of the gland itself.

Browne-Séquard has demonstrated that irritation of the skin at certain points causes epilepsy. Now, I submit that the phenomenon is produced in the organic nervous filament, which communicates with a sensitive nerve, and thus, in some instances, conveyed directly to the brain; in others, to the spinal cord and thence to the brain. The irritation thus propagated being extended to the central ganglion, and, by reflex action, from the ganglia to the brain and spinal cord; thence to the ganglia at the roots of the nerves; and, finally, to the nervous filaments supplying the muscles. Here the brain is known to be implicated by loss of volition, and the central ganglion and other ganglia by the convulsions which supervene. When a person dies of epilepsy, it often happens no organic lesion can be found; under such circumstances death being caused by irritation, and ultimately, destruction of the organic nervous system.

In delirium tremens, the irritation is propagated from the organic nerves in the coats of the stomach, by the par vagum, to the brain; from the latter to the central ganglion; by reflex action, to the brain and spinal cord, to the intervertebral ganglia and organic filaments. Hence the delirium can be explained—the nervous twitchings of the muscles, the convulsions as well as the death of the patient—by the irritation and final exhaustion of the organic nervous system: here, too, post-mortem examination often fails to account for death. Thus, a man apparently and in truth possessed of great muscular strength, dies in an incredibly short time; perhaps by an attack of convulsions or syncope; thus showing the cessation of the heart's action depends on the destruction of the organic nervous system.

Every physician knows that hysteria can be almost always traced to irritation of the genital organs. The organic nerves of the uterus act consentaneously with the central organic ganglion through the connection of the hypogastric and sacral nerves with the uterine and vaginal nerves to the spinal cord, and thence to the brain; the phenomenon which succeeds does not now require explanation.

Epilepsy, produced by masturbation of the genital organs, can, in the male, be readily understood in the same way. Compression of the brain, followed by loss of volition, sensation, and motion, as well as characterized by a slow pulse, stertorous breathing, and dilated pupils; here, recollecting the location of the central organic ganglion, and knowing that it can be compressed through pressure on the brain, it accounts for the condition of the patient.

A slice of the brain may be removed without disturbing the functions of organic life; because there is no pressure exercised on the central ganglion.

A man eats certain kinds of shell or putrid fish, and it is followed by cutaneous eruption; the connection here, between the organic nerves of the skin and the stomach, is the cause, by continuity of surface.

When a man gets concussion of the brain, he gets into a partial state of coma, from which he can be partly made conscious; but it is well known that, even when his intellectual faculties are almost totally annihilated, he will get up and pass water; this fact shows how the organic nervous system presides over the protection of the body. In making post-mortem examination of persons who have died of concussion of the brain, no lesion of the brain, very often, can be discovered. Mr. Colles used to remark in his lectures, that the only thing that could be observable was, that the brain seemed to be compressed so as not to fill the cranium.

Here, it will be perceived, if the brain is thus circumstanced, it must necessarily compress the pineal gland, or central ganglion of organic life, and thus squeeze the vital principle out of the gland, and, consequently, the destruction of life itself.

In poisoning from lead, the muscular fibres of the intestines become spasmodically contracted, although not furnished with motor-spinal nerves, or apparently any nerves. The abdominal muscles become rigidly contracted after some time, showing that the same influence operates over them. If the irritation is kept up, paralysis of the muscles follows, particularly in the arm, showing the destruction of the organic nerves.

In caries of the vertebræ, the muscles of the lower extremity are rigid, and a man walks as if on stilts. Here the irritation is propagated to the intervertebral ganglia and organic filaments, proceeding from the latter to the muscles.

A man gets bad typhus fever; the abdomen becomes tympanitic; the pulse scarcely perceptible; the action of the heart extremely feeble; with involuntary discharges of fæces and retention of urine. Here the organic nervous system is on the brink of death. It has lost its power over the intestines and heart, and it cannot keep garrison over the bladder or rectum.

A man gets ileus, or invagination of the intestines; is followed by spasms of the muscular fibres of the part affected; the part of the intestines above the stricture becomes dilated, whilst the part below is constricted. When mortification sets in, the powers of life situated in the organic system quickly give way; recognized by the absence of the pulse; the intermitting action of the heart; the cold, clammy perspiration; the hiccup and hippocratic countenance, together with the relaxation of the bowels; thus showing that death has destroyed the barrier of obstruction.

When a man wishes to feel anything, he directs the mind to it; communication is thus had with the central ganglion, which acts by reflex action through the brain, spinal cord, and roots

4

of the spinal nerves on intervertebral ganglia and organic filaments accompanying the brachial plexus; at the top of the index finger the branch of the median nerve and organic nerve inosculate, and thus the mind is made sensible of the nature of the part touched. Thus, the organic nerve acts as an afferent and efferent nerve.

When a young fellow sees a handsome girl, he becomes enamored with her; the genital organs sympathize; here the optic nerve communicates the impression to the brain; the latter corresponds with the central ganglion, which, by reflex action through the brain and spinal cord to the sacral ganglia and plexus, the spermatic nerve inosculates with the organic nerves in the organs of generation, and here they act and harmonize with the brain.

John Hunter remarks, to perform the act of copulation well, the mind must be fully intent on the object. This proves that the brain must act through the central ganglion.

It therefore follows that the organic nerves act at all times consentaneously. I am satisfied the nerves in the fœtal and maternal vessel form a ganglion in the placental lobule, thus keeping direct communication up between mother and child, and thus verifying the truth of the Gospel: "And it came to pass that when Elizabeth heard the salutation of Mary, the infant leaped in her womb."

When a man gets an attack of intermitting fever, the rigor is characteristic; the organic nervous system is all out of order, and the condition of the surface shows how much the organic nerves in the skin are implicated. Reflection will at once suggest what occurs.

Here let me observe, that God showed, by his own act, that air or breath was necessary for the life of man; for the moment breathing ceases, man is the same as when God formed him of the slime of the earth. Man is made of the elements of the earth, and how true it is: "Remember, man, thou art but dust, and into dust thou must return."

Having now demonstrated that the iris acts under the influence of the central ganglion, I am anxious to point out some practical hints: every person knows, when opium is taken in small quantities, it produces exhilaration of spirits, exciting the cerebro-spinal nervous system; but when taken in excess, it stimulates the organic system to such a degree as to induce fatal consequences; this fact is proved by the contraction of the pupils to a mere point. Drs. Corrigan and Graves have shown that belladonna and opium are mutually remedial, and, in the present number of *Braithwaite's Retrospect*, Dr. Benjamin Bell alludes to a case of poisoning by muriate of morphine, when belladonna had the effect of neutralizing the pernicious effects of the morphine. The belladonna has decidedly a sedative action on the central ganglion, as shown by the dilated pupil, so that its *modus operandi* can be readily understood.

In tetanus, chloroform should be administered, together with supporting the system. Chloroform acts as a sedative on the organic nerves, and relaxes the muscles. Here I should state, I reported a case of traumatic tetanus, a few years ago, in *Reese's Medical Gazette*, in which the treatment just mentioned was attended with the happiest results, namely, the recovery of the patient.

The late Dr. Graves, in speaking of the treatment of typhus fever, said he wished to have inscribed over his tomb, "that he fed fever." The stimulants and restoratives enrich and enliven the blood, which acts on the organic nerves, lining the bloodvessels, as well as the entire ganglionic system.

There is a case lately reported in the *Lancet*, where Dr. O'Reilly, of St. Louis, treated a case of poisoning by strychnine, by nicotine, with success. This case just proves, that as the strychnine causes spasms of the muscles, by its irritant effect on the organic nervous system, that nicotine produces relaxation of the muscles, by its sedative action.

In puerperal convulsions, the organic nerves in the walls of the uterus (the branches of the hypogastric and sacral plexus) inosculate with the uterine and vaginal nerves; the disturbance is propagated to the spinal cord by the latter, to the brain and central ganglion, and by reflex action from the latter to the brain and all the nerves proceeding from the cerebro-spinal axis. That this is true, is witnessed by the frightful rolling of the eyes, showing the lenticular ganglion has no control over the muscles it supplies; by the grinding of the teeth; strangulation and convulsive movements of the muscles. Chloroform here should be the remedy to be relied on; and I find that Dr. R. J. Tracy, Physician to the Melbourne Lying-in Hospital, (see *Braithwaite's Retrospect*,) has administered it in cases of puerperal convulsions with the most satisfactory results.

In hysteria there is titillation, or pleasurable excitement created in the organic nerves of the uterus, which is conveyed by the usual messengers to the brain and central ganglion, and, by reflex action, from the latter to the brain and other ganglia, to the filament proceeding from the latter. Here it must be remarked, that a female, laboring under a paroxysm of hysteria, will be conscious of what is taking place; showing that the cerebrum—the seat of intellect—is not interfered with. Turpentine destroys the tickling sensation of the nerves, and hence one of the best remedies in an attack of hysteria is a large turpentine enema, as well as the administration of the same medicine internally. Dr. Elliotson well remarks, it has a specific effect when it reaches the colon.

When a man gets a lacerated and contused wound of one of the extremities, and it is followed by mortification, some surgeons maintain that amputation should not be performed until the line of demarcation takes place, whilst others advocate a different practice. It is evident the destruction of the soft parts is the effect of the death of the organic nerves in the part, and that the entire organic nervous system is suffering in consequence of a part of it dying; this is clearly evident from the symptoms. As every inch the mortification extends cuts off a link of life, on true physiological principles it would be better to remove the diseased part at once. It may be, and is objected, that if the limb is removed before the line of separation sets in, the stump will be attacked with mortification. When amputation is performed too near the dead part, there is no necessity for ligatures; the blood is found coagulated in the vessels; this clearly shows that the organic nerves in the internal coats of the blood-vessels are contaminated, and that, as a matter of course, a return of the mortification is to be expected.

Under such circumstances, therefore, amputation should be performed at a very long distance from the diseased part.

Should chloroform be given to a person whose organic nervous system is so smitten? I think not. The vital power situated in the organic nervous system is almost extinct, and ready to depart at a moment's notice: chloroform, by its sedative influence, would probably annihilate it *in toto.* In such a case, French brandy and other stimulants should supplant chloroform, and prop up the drooping powers of life.

A man who falls from a height, on the top of his head, will be found laboring under symptoms of concussion of the brain; blood will be found flowing from his ears, and after some time it will cease and be replaced with serum—a certain sign, according to the late Drs. Dease and Colles, that fracture of the base of the skull has taken place; the man dies; extensive extravasation of blood may be found under the cerebellum. Why, it may be asked, are there not symptoms of compression under such circumstances ? The answer is, because the central ganglion is not compressed, the tentorium cerebelli acts as its safeguard.

It seems strange how one man may get a depressed fracture, followed by all the symptoms of compression of the brain, and another may receive, apparently, a much greater injury of the skull without such complication. The explanation depends on the part injured, as well as the region of the skull; if pressure is not made on the ganglion, the symptoms are only indications of concussion.

It is a remarkable fact, that a man who has got concussion of the brain will have a pulse ranging between 80 and 100, as long as he remains lying on his back; but on making him set up, the pulse will rise to 120. How is this to be accounted for? When lying on the back there is more blood sent to the brain, as well as to the central ganglion; consequently, the latter has more power over the heart, through the par vagum and cardiac plexus. The principle that operates here is the same that guides the treatment of syncope when the person is placed on his back. When a person is exhausted from immense hæmorrhage, as parturient women sometimes are, every person is aware that getting he patient to sit up in bed may cause instant death. The heart is a vital organ; it is presided over by a vital organ, and that vital organ is the central ganglion located within the head; and that it is so is manifested, inasmuch as it ceases to exist the moment the stimulus of the blood is withdrawn from it, by the patient assuming the erect position. When the organic ganglia lose the stimulus of the blood, they suffer from irritation and exhaustion; hence the convulsions and death which ensue.

Sir Astley Cooper cautions surgeons not to have recourse to blood-letting in persons who have fallen on race-courses, as by doing so they abstract the natural stimulus for the heart's action. This is quite correct; but it is clear it goes further—it prevents the heart from sending the requisite stimulus to the central ganglion. Here, again, I must reiterate, the brain has no influence over the action of the heart, as is proved by the vigorous state of the circulation in acephalous monsters.

Any oversight that I have been guilty of in writing this paper I hope will be pardoned, when I state it was written under a heavy press of professional business, *currente calamo*.

Connection of the Nervous Centres of Animal and Organic Life, with the Results of Vivisection.

My presence being casually requested, in the middle of November last, to take a part in a discussion relative to the anatomy and physiology of the placenta, it so happened on that occasion, that I was induced to advance original views in the course of the debate on the matter which took place.

In order to sustain the position I assumed relative to the nervous connection between the mother and fœtus, it became necessary to study the action of the nervous systems of animal and organic life. I state these particulars for the purpose of showing I had not directed my attention to these important subjects previously, and as an apology for the abrupt manner in which the papers were written.

To promulgate new doctrine with a view of overthrowing the old, unless founded on unquestionable facts, must, to a great extent, be deemed a heresy.

No person entertains a higher opinion of the labors of Prochaska, Bichat, Richerand, Sir Charles Bell, Payne, Hall, Müller, Brown-Séquard, Bernard, Grainger, Todd, and Dalton, than I do. I trust, therefore, I will not be considered as detracting from the reputation of these distinguished physiologists, when I assert they have failed to fully or truly expound the relationship existing between the nervous systems of animal and organic life.

It will be recollected, I asserted that the lenticular ganglion acted under the control of the Pineal gland, or ganglion, through reflex action through the brain, and the *exceedingly* small filament of the third nerve communicating with it; and that the contraction and dilatations of the iris were thus regulated, as well as the adjustment of the muscles of the eyeball, supplied by the third pair of nerves; that the par vagum acted under the influence of the central ganglion; that tetanic spasm of the muscles was produced by irritation of the organic ganglia; that coma was induced by compression of the central ganglion.

In a conversation I recently had with my esteemed friend, Dr. Busteed, Professor of Anatomy and Physiology at the New Veterinary College, New York, who is not only a capital anatomist, but an excellent operating surgeon, and one to whom I am greatly indebted for kindly volunteering to test my theory by vivisection, I stated my conviction, that the carotid plexus governed the circulation of the blood, and prevented the brain from being broken up, or deluged with blood, when attacked with inflammation, as well as protected the lives of persons indulging to a fearful and amazing extent in the imbibition of intoxicating drinks.

As it may be deemed superfluous to make further allusions to the statement made in my former paper, I beg to refer to the essay itself.

In order to afford facility to others to make experiments for their own satisfaction, I will state the mode of proceeding: The best animal that can be selected is a sheep; none other can be kept sufficiently quiet. The instruments required consist of a hand-saw, chisel, dissecting-knife, forceps, and retractors; sponges, water, ligatures, and plugs of paper about a quarter of an inch in thickness should be in readiness. The sheep, being placed on a firm table, should be held by assistants. The scalp is now to be freely removed, together with the muscles attached to the cranium. A coffin-shaped piece of the skull is next to be taken away, about four inches long and three in width; the narrow extremity should terminate at a line drawn transversely about one inch and a half above the superciliary ridges, whilst the posterior should extend about one inch beyond the occipital protuberance. This being accomplished, the dura mater is next to be dissected off by dividing it on each side of the falx, separating it from its attachment to the crista galli; here some hæmorrhage will take place from the longitudinal sinuses; but it need not be apprehended. The membrane being reflected back, the tentorium cerebelli is next to be detached. In doing this, the lateral sinuses will be opened, rendered manifest by a gush of venous blood, which, if not restrained, will soon render further dissection useless; several sheep were prematurely killed by this accident; the top of the finger should be pressed against the part until a plug of paper is pushed into the sinus, which will prevent further trouble from this source. Having now exposed the cerebrum and cerebellum, the posterior lobes of the cerebrum should be separated from the cerebellum. In doing this, there will be some hæmorrhage, which may be restrained by the application of ligatures. Having reached the fissure of Bichat, the posterior border of the corpus callosum, together with that portion of the fornix incorporated with it, must be divided in the mesial line from before backward, and held asunder by retractors. The velum interpositum is now exposed; it must be divided in the same direction. Hæmorrhage to some extent will be the result, which may be arrested by the application of cold water. The venæ Galleni, which carry blood from the plexus choroides to the straight sinus, are inclosed in a duplicature, divided from the velum interpositum, and cannot escape being derived. The tubercula quadragemina are now brought into view. A small pale, yellowish-red body will be seen anteriorly resting against the nates, connected to the optic thalami at the sides, and placed just above the iter a tertio ad quartum ventriculum; this is the Pineal gland or ganglion. It will be perceived, in the latter part of the dissection described, which is very difficult to perform, in consequence of the oozing of blood, that every time the point of the knife goes into the neighborhood of the gland, the sheep plunges.

The gland being now open to observation, you gently seize its body with the forceps, extending the points towards, or nearly as far as its attachment to the thalami. Oscillations of the iris will be at once the effect. Move the gland more freely, and the pupil will contract to a very small diameter in an instant. Still further press the gland, and make traction, and the eyeball will move rapidly in all directions.

Here let me anticipate a question that may be asked, namely: How do you know but the fourth nerve, which goes to the superior oblique muscle, and the sixth, which is distributed to the external rectus, is not engaged in these movements of the eye-ball?

My answer is, it would be all nonsense to suppose so, inasmuch as the fourth nerve arises by three filaments from the valve of Vieussens, and the sixth from the superior extremities of the pyramidal bodies, and are several leagues distant, figuratively speaking, from the Pineal gland, and have no connection with the lenticular ganglion.

Make more firm pressure and traction, and the sheep will vomit and be thrown into a tetanic spasm; the neck will be curved, the legs thrust violently forward, every muscle in the body will appear to be engaged, and you will hear the by-standers exclaim, "You have killed the sheep!" The forceps being now relaxed, the sheep will shortly recover, and you can go through the same process. Sometimes, instead of the vomiting, the sheep will bleat most pitifully, and then be thrown into the tetanic spasm. Press the gland down towards its attachment, and you will observe the pupil to dilate.

Another way may be taken to expose the gland, which consists in separating the hemispheres of the cerebrum, slicing off the lobes on a level with the corpus callosum; in doing which it is probable a small opening may be made into the lateral ventricles, which will be soon known to have taken place by the sheep commencing to snore, in consequence of the blood passing into the ventricles and compressing the gland. The corpus callosum is next to be reflected back. The anterior pillars of the fornix are next to be detached and thrown back as far as possible, which is rather difficult, in consequence of its connection posteriorly with the corpus callosum.

The clot of blood being removed, the gland will present itself, and the sheep will breathe naturally. The brain may be cut in all directions and freely removed, without apparently producing the slightest effect on the sheep. It may be pressed on with same indications. Not a single vessel will be observed to give blood *per saltum*, notwithstanding the deepness of the incisions; and the animal suffers no pain so long as the gland, or the part of the brain the gland is attached to, is not interfered with.

Having now described the operations as performed by Dr. Busteed, I will proceed to state the result of the experiments:

First Sheep.—After exposing the brain, Dr. Busteed passed in a very fine needle, with a view of piercing the gland; the sheep was instantly attacked with violent convulsions, which continued until the butcher cut the sheep's throat. The brain was now removed, and the needle was found to have passed through the peduncle of the gland.

Second Sheep.—After exposing the gland, compression produced dilatation of the pupil and tetanic spasm, when the sheep's throat was cut.

Third Sheep.—Hæmorrhage rendered the operation unsatisfactory.

Fourth and fifth Sheep.—Contraction, dilatation, rolling of the eyeballs, and tetanic spasms.

Sixth Sheep.—Similar results, together with vomiting, on the gland being drawn from its attachment, which was instantly followed by tetanic spasm and death of the sheep.

Seventh and eighth Sheep.—Contraction and dilatation of pupil; rolling of the eyeballs. Pressure with traction caused the sheep to bleat most pitifully, as if suffering extreme torture, followed by tetanic spasms. In these two cases the experiment was performed in the manner secondly described, and the sheep were observed to commence snoring. After the hemispheres of the brain were removed in both cases, the bodies of the lateral ventricles were slightly opened, allowing the blood to flow in. On removing the fornix and the clot of blood, the sheep breathed naturally. Supposing the seventh sheep had died in the tetanic spasm, the butcher cut off the head, when the body of the sheep plunged violently, and life was not extinct for about three minutes.

Ninth Sheep.—Contraction and dilatation of pupil, rolling of the eyeballs, tetanic spasms, and death. Here let me observe, the bleating of the sheep was caused by the irritation propagated through the recurrent branch of the par vagum, to the organic nerves in the larynx, derived from the superior cervical ganglion.

I will now endeavor to demonstrate that the great Sir Charles Bell did not thoroughly understand the use of the spinal nerves. I will quote Sir Charles's remarks. He says:

" On laying bare the roots of the spinal nerves, I found that I could cut across the posterior fasciculus of nerves, which took its origin from the posterior portion of the spinal marrow, without convulsing the muscles of the back; but that on touching the anterior fasciculus with the point of the knife, the muscles of the back were immediately convulsed."

Here let me remark, the anterior and posterior roots of the spinal nerves, as Mr. Grainger has beautifully shown, receive filaments from the pre-vertebral ganglia, and consequently the anterior roots of the spinal nerves could not be irritated without touching the filaments of the ganglia, thus showing the fallacy of the experiment.

The spinal nerves are messengers, instituted to carry instructions relative to sensation and motion, from the nervous centre of animal life to the nerves of organic life, situated at remote parts of the body.

If all the muscles in the body can be thrown into spasmodic action by irritating one ganglion in the brain, it follows, as a consequence, that the spinal nerves are not actually required for that purpose.

Comparative anatomy shows that the spinal nerves are not essentially required for the movement of the muscles. The invertebrata have no spinal cord or spinal nerves; yet the snail slowly winds his way with his domicile on his back, and if he happen to be decapitated during his perambulations, he is not much discomfited, but will soon reappear with a new head. The little busy ant moves with astonishing rapidity and enjoys good visual organs, thus showing the perfection of its nervous system. I cannot help remarking, it would take an admirable anatomist to prepare the retina of the ant for demonstration; as well as a powerful microscope to bring into view the foramen of Sœmmering. I am induced to make these remarks to show persons, who will not believe anything unless revealed by the microscope, that nerves exist where they are not able to discover them.

Dr. Hall states the real objects of his researches as follows:

First.—To separate the reflex actions from any movements resulting from sensation and volition.

Secondly.—To trace these actions to an acknowledged source, or principle of action in the animal economy—the vis nervosa of Haller acting according to newly discovered laws.

Thirdly.—To limit these actions to the true spinal marrow, with its appropriate incident and reflex nerves, exclusively of the cerebral and ganglionic systems.

Fourthly.—To apply the principle of action involved in these facts to physiology, viz.: to the physiology of all the acts of exclusion, of ingestion, of retention, and of expulsion in the animal frame.

Fifthly.—To trace this principle of action in its relation to pathology, viz.: to the pathology of the entire class of spasmodic diseases. And,

Sixthly.—To show its relation to therapeutics, especially to the action of certain remedial, and certain deleterious physical agents.

Finally.—To these objects, taken together as a whole, or as a system, I prefer my claim, and I do not pretend that an occasional remark may not have been incidentally made by some previous writer bearing upon some one or other of them.

It does not come within my province to discuss whether Prochaska, Payne, Hall, or Campbell is entitled to the honor of the discovery alluded to by Marshall Hall; but I affirm, inasmuch as spasm of the muscles can be induced by irritating the Pineal gland, or central ganglion, that the discovery, no matter who claims it, is shaken to its very foundation, and thrown completely in the background, inasmuch as a large mountain, called the Pons Varolii, separates the Pineal gland from the medulla oblongata. It is not, therefore, necessary to prove that Marshall Hall arrived at false deductions, having drawn them from false premises.

Dr. Brown-Séquard has shown that certain parts of the body are susceptible of producing epileptic attacks on being irritated; and has demonstrated the fact by tickling a monkey under the angle of the eye, and thus inducing epilepsy. Here let me remark, that monkeys are prone to very vicious habits, and are continually practicing masturbation—thus rendering their organic nervous system in the highest degree excitable, and prone to epileptic seizures. Every person knows that tickling the soles of the feet will produce convulsive attacks; and if continued, will end in death. The phenomena just enumerated are considered the result of reflex action, and excitation of excito-motor, or incidental nerves.

It is a certain fact that the application of belladonna to the eyelids causes dilatation of the pupil; it is equally true that the iris is furnished exclusively with nerves from the lenticular ganglion. How can the matter be accounted for by the explanation of either Brown-Séquard or Marshall Hall? I would answer, it cannot. What occurs is this—the belladonna exercises a sedative action on the organic nerves in the skin, which inosculate with the nasal branch of the fifth pair; and is conveyed by the latter to the brain, central ganglion, and by reflex action from the ganglion to the brain, fifth nerve, ophthalmic division of the fifth, nasal branch of the latter, and lenticular ganglion by its communicating branch, and thence to the ciliary nerves that go to the iris. This is a most beautiful illustration of reflex action according to my theory, and solves a difficulty that I believe no man has satisfactorily done before.

When a young lady is spoken to about her lover, she instantly blushes. How is this to be accounted for? The mind communicates with the central ganglion, the latter by reflex action through the brain and facial nerve to the organic nerves in the face, with which its branches inosculate. This may be said to be a fanciful theory; but does not the stomach blush during the process of digestion? Is it not supplied by branches from the solar plexus which inosculate with the branches of the par vagum? Here I must state, it appears from the experiments of Bernard, as well as the examples just given, that the blood becomes oxygenized whenever the animal and organic nerves are acting in conjunction, and confirms the doctrine I advanced on a former occasion, that the fœtal blood was arterialized by nervous influence in the placental lobule. Marshall Hall gives us an example of reflex action: the grasping of the ovary by the fimbriated extremity of the Fallopian tube during impregnation. The fœtus of the kangaroo, after leaving the uterus, becomes attached to the nipple of the dam. The infant, after being born, seizes the nipple of the mother in its mouth. Are not these examples analogous to the one pointed out by Marshall Hall? I think they are, and at the same time think it would be a very difficult thing to demonstrate nerves passing from the lips of the fœtus to the true spinal marrow of the dam in the one case, or the lips of the infant to the mother in the other. How can these things be explained? The answer is—by vital action. What is vital action? It is an intelligent power, inherent in and emanating from the organic nervous system. It is the breath of life, which, when blown out, terminates man's earthly career, and leaves him an inanimate mass—such as he was when God made him out of the slime of the earth. It is what is generally known as instinct. How can this be proved? By recollecting the part the lenticular ganglion plays in the regulation of the movements of the iris, and the otic in the tensor tympani muscle, as I have already described in my former paper.

I am satisfied many persons will deem me very presumptuous for controverting the opinions of Marshall Hall; but my aim being truth, I am regardless of criticism. "*Magna est veritas et prevalebit.*"

I think I have sufficiently elucidated my theory to enable any person to understand the pathology of catalepsy, chorea, paralysis agitans, spasm of the glottis, convulsions from dentition, spasmodic asthma, sea-sickness, and other diseases too numerous to specify.

When persons are exposed to the vapors of certain gases they are said to die of apnœa, from *a* = *non*, πνεω = *spiro;* which translated means, in popular language, "shortness of breath," or asphyxia, from *a* = *non*, σφυξη = *pulsus*, which may be literally translated, the stroke of death. What causes death under such circumstances? The blood becomes charged with carbonic acid, and acts as a poison on the organic ganglia, which are thus rendered powerless in their function of causing the muscles to contract; hence it is the heart immediately ceases to pulsate, and the consequent death of the individual.

The mode in which arsenic, tartar-emetic, tobacco, and other poisons destroy life, is by their destructive influence on the organic nervous system.

As some may suppose I am stating things not called for in an article of this kind, I will conclude by observing, the experiments detailed were performed by Dr. Busteed in presence of Dr. Gallagher, Mr. Roderick, myself, and several other persons. I have further to remark, that I will leave it to others to decide whether I have made a discovery or not, with respect to the anatomy and physiology of the placenta, or the connection of the nervous centres of animal and organic life.

If I have appropriated the offspring of any other man's brain to myself, I candidly acknowledge I have done so in blissful ignorance of the name of the author; and will make a suitable acknowledgment on being apprised of the name of the individual, and where his work is to be found.

If my former communications commanded no attention, I attribute it to my not being known as a teacher, or being connected either with a college, a hospital, or any other public institution. I do not feel surprised, being thus circumstanced, that many should think I could have no knowledge of the subject I attempted to elucidate. Everybody knows a professor is looked upon as being something extraordinary, and that it may be said of him—

"And still they gazed, and still the wonder grew,
That one small head should carry all he knew!"

NOTE.

The following extract, taken from the third volume of the Cyclopædia of Anatomy and Physiology: Article, Nervous Centres, by R. B. Todd, is worthy of attentive perusal, as showing the *important connections* of the ganglion, as well as the opinion entertained of its use by Des Cartes, who, it is almost unnecessary to state, considered the gland the seat of the soul, without being able to demonstrate the fact, or show any connection between *it* and the other ganglia. And here let me state, that this latter observation I believe to be true of every writer since Des Cartes wrote his work; and that my discovery consists in showing the Pineal gland is the chief ganglion, and that it is connected with all the other ganglia, and presides over them.

"PINEAL GLAND.—We may here conveniently notice the position and connection of the *Pineal gland*. This body, rendered famous by the vague theory of Des Cartes, who viewed it as the chief source of nervous power, is placed just behind the third ventricle, resting in a superficial groove, which passes along the median line between the corpora quadrigemina. It is heart-shaped, and of a gray color. Its apex is directed backward and downward, and its base forward and upward. A process of the deep layer of the velum interpositum envelops it, and serves to retain it in its place. From each angle of its base there passes off a *band* of *white matter*, which adheres to the inner surface of each optic thalamus. These processes serve to connect the Pineal body to the optic thalami. They are called *the peduncles of the Pineal gland*, also *habenæ*. In general they are *two* in number, one for each optic thalamus. They may be traced forward as far as the *anterior pillars* of the *fornix*. Posteriorly these processes are connected along the *median line* by *some white fibres* which adhere to *the base* of the *Pineal gland*, as well as to the *posterior commissure* beneath, and which seem *to form part* of the *system* of *fibres* belonging to that commissure. A *pair* of *small bands* sometimes pass off from these fibres, along the optic thalami, parallel to the peduncles above *described*."

✸

5

Further Remarks upon the Connection of the Nervous Centres of Animal and Organic Life.

Every operating surgeon knows, to perform an operation on the dead subject, is quite a different thing to doing it on the living body.

The same remark is equally true with respect to exposing the Pineal gland in the living or dead animal. In the former there is great trouble; in the latter there is no difficulty.

By a simple experiment, any person can satisfy himself that the Pineal gland is a ganglion of the organic nervous system, as piercing the gland with a fine needle will produce the following phenomena:

Oscillation of the iris, contraction of the iris, rolling and fixing of the eyeballs, and tetanic spasm of the muscles of the body.

It should be observed, unless the gland is touched, no such effects are produced by puncturing any other part; which fact affords in itself, in the strongest manner, negative proof of the importance of the ganglion.

A glance at the annexed plates, drawn by my friend, Mr. William Henessy, from dissections made by myself, will show the mode of carrying out the vivisection, and demonstrate the relative anatomy of the ganglion.

A perusal of the first and latter part of my paper, to which I most earnestly and respectfully beg to refer, will convince any person of the truth of the deductions to be arrived at from what takes place when the ganglion is irritated.

If one line be drawn transversely, so as to allow the posterior lobes of the cerebrum to touch it, and another in the direction of the longitudinal fissure, the ganglion will be found in the median line, at a distance of three-quarters of an inch from the transverse, and at a depth of one inch and an eighth from the peripheral surface of the cerebrum.

In case an attempt is made to puncture the ganglion, and that it escape being wounded, the sheep will fall into a state of coma, and commence snoring.

If the gland is now cut down on, it will be found surrounded by a clot of blood.

It is a remarkable fact, that life should exist not only in the body of the sheep after decapitation at the articulation between the atlas and occipital bone, but likewise in the head, notwithstanding the mutilation of the brain, and the destruction of the Pineal gland or ganglion. Yet such is the case; and the mouth will shut and open once or twice. Here it will be recollected that the vital agent is in the spheno-palatine ganglion.

It will be recollected I maintained, throughout my observations, that each organic ganglion was the seat of vitality. If an eel be cut in half a dozen pieces, each part will be alive, because each part contains one or more organic ganglia.

Nutrition, assimilation, secretion, and absorption are the result of organic nervous influence. How are these important matters provided for in the encephalon? Where is the organic ganglion to be found, destined to preside over these functions in the brain? I will answer, In the *sella turcica*, and is that body called the pituitary gland, which is composed of gray and white matter, incased in the *dura mater;* enveloped by the arachnoid membrane, communicating, through the *infundibulum,* with the third, as well as all the other ventricles of the brain, and, by the continuity of surface of the arachnoid membrane, with the entire surface of the cerebral mass. I should also observe, that it appears to be in direct communication with the *Pineal* gland, or central ganglion; that the pedunculi of the gland can be traced down towards the *infundibulum.* (See plate.) It may be said no nerves can be detected in the arachnoid membrane; but the same objection holds good with respect to the *pericardium, pleura,* and peritoneum. However, when these membranes, as well as the arachnoid, are in a state of inflammation, the exquisite pain proves, beyond a doubt, the existence of nerves; the vascularity shows the presence of blood-vessels, although such could not be previously discovered; and the effusion of lymph, serum, or pus, demonstrates that secretion is vigorously carried on, and the subsequent removal of these substances points out the activity of the absorbents.

It will be perceived the semilunar ganglia perform in the abdomen the same kind of duties the pituitary gland or ganglion does in the cranium. These ganglia further resemble one another in being located in secure positions, being in the proximity of large blood-vessels; in being at some distance from the organs they supply with nerves. The *white* bands constituting the pedunculi of the Pineal gland, as before stated, can be seen proceeding towards the *infundibulum*, which passes down from the third ventricle to the pituitary gland; the internal carotid arteries pass by the sides of the pituitary gland, surrounded by a plexus of nerves derived from the superior cervical ganglia, if any of the branches of the plexus enter the gland, and I am certain they do: so, then, a complete communication would be established between all the ganglia in question.

I will not trespass for some time on the readers of the GAZETTE. I therefore expect whatever blunders I have committed, (and they are not a few,) or presumption I may be deemed guilty of, will be magnanimously pardoned by all, not excepting those who are inclined to split hairs, and who regard everything savoring of originality *aduncis nasis.*

FIG. 1.—Appearance of Cerebrum and Cerebellum after the removal of the Calvarium and Dura Mater.

1, 1, Cerebrum.   2, 2, Cerebellum.

FIG. 2.—Similar View as Fig. 1, with rule and needle describing the method for finding the Pineal Gland or Ganglion.

FIG. 3.—Hemispheres of the Cerebrum removed on a level with the Corpus Callosum. The posterior lobes of the brain drawn upward and forward with the Corpus Callosum, so as to bring into view the Pineal Gland and its peduncles.

1, 1, Cerebrum.   4, Pineal Gland.
2, 2, Corpus Callosum.   5, 5, Nates.
3, 3, Pedunculi Pineal Gland.   6, 6, Testes.

Fig. 4.—Vertical Section of the Brain, showing the anatomical relations of the Pineal Gland.

1, 1, 1, Cerebrum.	6, Nates.
2, Corpus Callosum.	7, Testes.
3, Fornix.	8, Cerebellum.
4, Pineal Gland.	9, Fourth Ventricle.
5, Opticus Thalamus.	10, Medulla Oblongata.

Fig. 5.—View of the Base of the Brain, showing the Position of the Infundibulum.

1, 1. Cerebrum.	7, Medulla Oblongata.
2, 2, Olfactory Nerves.	8, Fifth Pair of Nerves.
3, Optic Commissure.	9, Facial Nerve.
4, Infundibulum.	10, 10, Sixth Pair of Nerves.
5, 5, Third Pair of Nerves.	11, 11, Ninth Pair of Nerves.
6, Pons Varolii.	12, 12, Fourth Pair of Nerves.

Fig. 6.—Horizontal Section of the Brain.

1, 1, Cerebrum.	8, 8, Plexus Coroides.
2, Anterior Commissure.	9, 9, Nates.
3, 3, Corpus Striatum.	10, 10, Testes.
4, Infundibulum.	11, Valvi Lucens.
5, 5, Pedunculi of Pineal Gland.	12, Medulla Oblongata.
6, Pineal Gland.	A, A, A, Lateral Ventricle.
7, 7, Optici Thalamici.	B, Third Ventricle.

Fig. 7.—View of the Base of the Brain, showing the Connection of the Infundibulum with the Pituitary Gland, after its removal from the Sella Turcica.

1, Pituitary Gland.	2. Infundibulum.

Fig. 1.

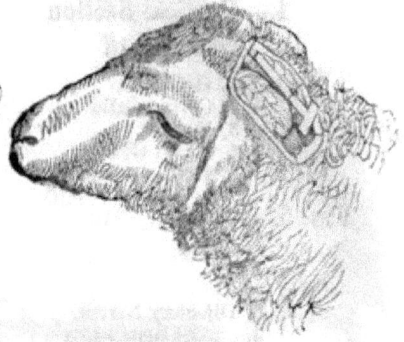

Fig. 2.

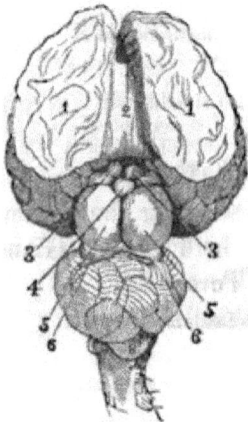

Fig. 3.

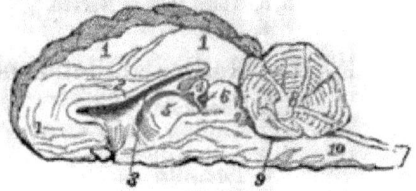

Fig. 4.

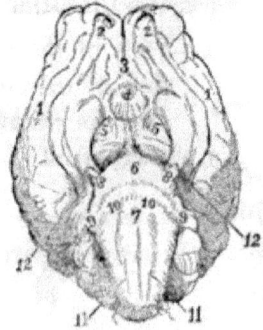

Fig. 5.

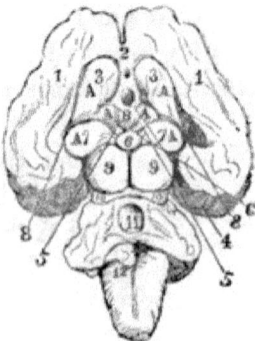

Fig. 6.

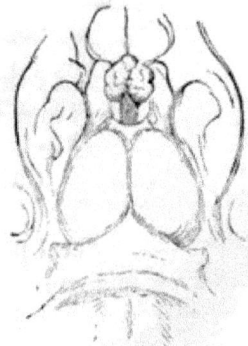

Fig. 7.

Observations on Syphilitic Iritis.

What is syphilitic iritis? It is a specific inflammation of the iris. What is its primary cause? An ulcer known as a chancre. How is the chancre produced? By impure sexual intercourse, whereby an animal poison is communicated from one individual to another. How soon is it known that inoculation has taken place? The time of the poisonous incubation varies from a few to several days, if the statements of individuals can be relied on. What are the characters of a chancre in the course of formation? Some heat, tingling pain, redness, a papilla followed by a vesicle; and lastly, a pustule attracts attention. What next happens? The cuticle cracks, leaving an ulcer or chancre exposed to view. What are the appearances of a chancre? I will answer by quoting Mr. Hunter's description: "The sore is somewhat of a circular form, excavated without granulations, with matter adhering to the surface, and with a thickened edge and base. This hardness or thickening is very circumscribed, not diffusing itself gradually and imperceptibly into the surrounding parts, but terminating rather abruptly." What is the *modus operandi* of the poison? The received opinion is, that it acts by absorption; in other words, that the poison is taken up by the lymphatics, conveyed into the circulation, and, after some time, makes its appearance at the site of inoculation.

Is there any objection to this mode of accounting for the phenomenon? Yes. If the poison acted by absorption, it is obvious that several parts of the body would be engaged at the same time; and consequently, that numerous chancres would be the result; and as such is not the case, it must be conceded that the poison is still located in its original situation—inasmuch as the chancre actually presents itself there. What, then, is the true explanation? This question must be answered by asking what is inflammation? A vast deal has been said and written about inflammation. Boerhaave attributed it to an *error loci;* Cullen, to spasm of the capillaries; John Hunter, to diminished muscular power, with increased elasticity of the arteries. Dr. Wilson Phillip remarked that the "circulation was slower in the capillaries." Dr. Thompson observed "increased action of capillaries in a moderate degree." Dr Hastings, " that inflammation consists in a weakened action of the capillaries, by which the equilibrium between the larger and smaller vessels is destroyed, and the latter become distended."

Are the phenomena attendant on inflammation now understood? The answer must be in the negative. It will be soon rendered apparent that the primary cause of inflammation is an impression made on the organic nerves: that when, for instance, the poison is brought in contact with the organic nerves, that they resist its pernicious influence for some time, but are eventually acted on by it; that they become excited, inducing or causing a dilatation of the arterial capillaries, as is witnessed by the redness and heat supervening.

Have the organic nerves the power to increase or diminish the calibre of the capilaries? Yes. The proofs appear to be irresistible. Is not a pale-faced man made to blush, or get a red face, when charged with an abominable crime? Does not a young, rosy-cheeked damsel grow deadly pallid when stricken with fear? Does not the mucous membrane or villous coat of the stomach become crimson and turgid during the process of digestion? Does not the penis, from being small and flaccid, become large and distended? Do not the lips and breasts become injected with blood whilst under excitement?

Is it not now clearly manifest that the organic nerves have the power to cause the expansion of the capillaries?

But, inasmuch as blushing only lasts for a few seconds, it may be objected that there is no similarity between it and inflammation. How is this difficulty to be got rid of? Easily—by recollecting the organic nerves, in the one case, are only in a temporary state of excitement; whilst in the other, the cause being kept up, the excitation is changed into stimulation.

How is it known that irritation has taken place? By the heat and pain which accompany it.

If the heat, pain, redness, and swelling depend on the dilatation and stimulation of the capillaries induced by the organic nerves whilst suffering from irritation, how is the effusion of lymph, serum, or pus, which subsequently occurs, accounted for? This interrogatory must be replied to by putting another. How are the gastric juice, the bile, and the urine secreted? Are not all these the product of the capillaries acting under the influence of the organic nervous system? Admitting the truth of this explanation, then, it may be averred that, precisely on the same principle, the lymph, serum, and pus are secreted from the capillary arteries whilst acting under the direction of the organic nerves.

Are there any other collateral proofs to sustain the declaration that the organic nerves are concerned, in the manner stated, in a case of inflammation?

Certainly. When belladonna is applied to the breast, or over the stomach, it causes dilatation of the pupil—the iris, it will be remembered, is exclusively supplied with nerves from the lenticular, (1) which is an organic ganglion; and is not influenced by the nerves derived from the animal nervous system. Again, the application of belladonna stops the secretion of milk —thus showing it has a sedative influence on the organic nerves.

If a blister is applied, and the cuticle subsequently removed, and strychnine is placed on the exposed surface, all the muscles will be thrown into tetanic spasm. But it may be said the strychnine is absorbed, and gets into the circulation. If such should be the case, then there would not be time for the muscles of the trunk to be thrown into spasm, as the heart, the great central muscle of the circulation, would be the first to suffer, and be attended with instant death.

When a blister is applied to the skin, it effects vesication as well as pain of the organic nerves in the rete mucosum. The kidneys also suffer from irritation, showing conclusively that the secretion of urine, as well as the exhalation from the skin, are under the same nervous influence.

When a person is attacked with constitutional disturbance

after a dissecting wound, a pustule will be found marking the situation of the wound, thus showing that the poison is still stationed at the point of its entrance.

An individual who has been bitten by a rabid dog, will have the wound healed up, and never after have his attention directed to it, until symptoms of hydrophobia present themselves, when the wound will be found exhibiting the signs of recent inflammation. What can be clearer than that the poison has been dormant at the place during the interval of the infliction of the bite and the ushering in of the symptoms of hydrophobia? A practical deduction should not be lost sight of—namely, excision of the part implicated, which may be put in practice, any time up to the inflammation of the wound setting in, with good effect.

The late Professor Colles, who was eminently practical, used to enforce in his lectures the propriety of keeping a patient under the influence of mercury as long as any hardness could be discovered, even although the chancre was healed up; stating, that as long as such continued, the patient was not in a position to escape secondary symptoms.

Here, too, the practical rule should be to exercise the part so affected. Here, too, it is evident the poison must be in its original nidus.

When a child is vaccinated, the vesicle and subsequent pustule appear where the matter was inserted. If the vaccine virus acted by absorption, then pustules would cover the whole body, and not be confined to one place.

When a person is attacked with a whitlow in the thumb, index, or middle finger, the radial artery pulsates violently, and demonstrates the energy of the organic nerves to overcome the difficulty opposed to them by the fibrous membrane. Further examples need not be multiplied.

Another question now suggests itself. What is the function of the capillary arteries when in a normal state?

To give oxygen to the organic nervous system, to provide for the wear and tear of the body; the restoration of organs acting under the influence of the organic nervous system, as is exemplified in the cicatrization of an ulcer. The blood contains the material required, and the organic nerves assimilate and form it to its proper mould.

Here I cannot avoid digressing to remark, that the pulmonary arteries, containing venous blood, terminate in capillaries in the bronchial membrane lining the air-cells; that the pulmonary veins, containing arterial blood, commence by capillaries in the air-cells; that the blood is separated from the air by a fine membrane; and consequently, that there must be some other influence to cause the absorption or endosmosis of the oxygen from the air, and the elimination of carbon from the blood.

What is the true explanation to be given? Again, analogy must be resorted to. When food is introduced into the stomach, gastric juice is secreted by the action of the branches of the solar plexus, (the stomachic plexus.) What food is to the stomach, air is to the lungs. The branches of the pulmonary plexus decompose the air, causing the oxygen to pass into the blood by endosmosis, and the carbon to be set free by exosmosis.

Again, on the surface of the body and other parts, the arteries terminate in capillaries, to give off the oxygen to the organic nerves; whilst the capillary veins carry back the blood, to be again revived. Here it will be observed the organic nerves have the power to take the oxygen from the air primarily, as well as to appropriate it to themselves subsequently. " Because the life of the flesh is in the blood."—(Leviticus, Chap. 17, verses 11, 14.) This is a divine truth. The blood contains oxygen, which is the life of the organic nervous system, without which life could not be retained.

Mr. Hunter's views with respect to the vitality of the blood are well worthy of consideration.

However, that life is not located in the blood, is proved by a simple experiment. When an animal is bled, and all the blood drained off, life will not be extinct; convulsions will take place in consequence of the struggle going on in the organic ganglia for oxygen.

Why death occurs on the abstraction of blood is now manifest; it is the want of oxygen. It is equally clear that whatever cause prevents the oxygenation of the blood, kills in the same way, inasmuch as life cannot continue in the organic ganglia without oxygen. In proportion as the quantity of oxygen is diminished, the foundation of life is undermined, as is witnessed in a case of phthisis, where the gradual destruction of the lungs cuts off by degrees the requisite supply of oxygen, until at length there is not enough left to prolong the flickering flame of life.

A clear and distinct knowledge of the organic and animal nervous systems is indispensably necessary for a proper elucidation of the subject under consideration.

Is life centred in the organic nervous system, consisting of organic ganglia and their branches; and is it perfectly independent of the animal nervous system, consisting of the brain, spinal cord, and spinal nerves?

It is certainly true that life inhabits, or is situated in, the organic nervous system, inasmuch as the brain may be totally destroyed without extinguishing life. I should remark, no man can demonstrate life—it resembles, in a certain degree, electricity—it is only known by its effects.

It is to be recollected that when man was made he was furnished with two sets of organs of the most beautiful, elaborate, ingenious, and complicated kind. The one intended for the maintenance and preservation of life, as well as the propagation of the species, namely: the organic nervous system, the organs of respiration, circulation, digestion, secretion, and absorption. The other, to connect the individual with the external world, consisting of the brain, spinal cord, spinal nerves, organs of sense and locomotion.

There are two intelligent, immaterial agents in man. One resides in the organic nervous system, and is possessed of extraordinary wisdom.

No man can make gastric juice, bile, urine, or saliva from the same material, or by any chemical process. If a man, ignorant of the fact, were told that the stomach, made up of muscular fibres, blood-vessels, nerves, mucous or villous membrane, and cellular tissue, could secrete a fluid susceptible of dissolving a piece of flint or steel, as the stomachs of some animals are capable of doing, he would deem it impossible. The giving of life to man was one act on the part of the Deity. See Genesis, Chap. 2, verse 7: "And breathed into his face the breath of life." The other in the animal nervous system.

The giving of judgment, volition, and memory, was an attribute conferred on man, independent of life, as is implied in the 2d chap., 17th verse of Genesis: "But of the Tree of Knowledge of good and evil thou shalt not eat."

The nerves of animal life are connected with the nerves of organic life. They act in concert, as occurs in the eye, the ear, and the larynx. To give an illustration: a singer wishes to produce a very high or low note; the will is conveyed by the recurrent branch of the par vagum to the muscles of the larynx. The *chordæ vocales* must be rendered tense or relaxed at the same moment by the action of the muscles. Now, the will has no power to produce the condition of the muscles desiderated; it therefore devolves on the organic nerves to regulate the muscles, which they do in the same way the ciliary nerves regulate the iris, and the otic nerves the tensor tympani muscles.

Here the combined wisdom of both intelligent agents is made conspicuous.

Here some person will exclaim, how is it possible a ganglion could possess such intelligence?

But it is just as difficult to conceive how a soft, pulpy mass, made up of white and gray substance, as is the brain, could contain all the knowledge of Sir Isaac Newton, or some such other learned man.

A philosopher may admire the mechanism and construction of the eye, as an optical apparatus, and contemplate with pleasure the image of a picture impinged on the retina; but let him be shown a section of the optic nerve, a white, opaque, soft cord, and be told it was susceptible and capable of carrying the description and appearance of all things in nature to the mind, he could not understand on what physical principle such an operation could take place.

Electricity cannot be made to pass through a wire until it is first generated; nor can the optic nerve discharge its functions, unless animated by the Spirit of Life.

I think this is now the proper place to inquire how the change in the blood in inflammation is produced.

I presume it is by the excited action of the organic nerves on the blood passing through the arterial capillaries.

I think I am correct in stating there is more oxygen in inflamed blood, than when no inflammation exists; that the nerves in the inflamed part over-stimulate the blood.

Here I should observe, that it is well known the blood of pregnant women presents the characteristic signs of inflammation. If my views are correct, the blood in the placenta is oxygenized for the fœtus by the action of the organic nerves, in what I have elsewhere called the placental lobule, on the same principle as inflammation.

Heat is one of the signs of inflammation. That it depends on excitation of the organic nervous system, I think can be incontrovertibly proved. When a person is attacked with intermittent fever, during the rigor, the respiration goes on, notwithstanding which the body, as well as the extremities, are deadly cold, even though it should be the hottest day in summer. When reaction sets in, no matter how cold the day may be, the surface of the body will become burning hot. In the former case, the organic nervous system is almost overpowered; but in the latter, on recovery from the shock, it is inspired with new energy. If animal heat depend on the combustion of oxygen and carbon in the lungs, how are the phenomena above alluded to to be explained? If animal heat depended on the combustion of carbon and oxygen in the lungs, how is the increased temperature in inflammation to be explained, where there is no air to procure the oxygen from, or carbon to combine with it? Again I will affirm, the immaterial, intelligent, vital agent that has its abode in the organic nervous system explains the difficulty.

Having now introduced what may de deemed extraneous and irrelevant matter, I will proceed to the examination of what takes place after the chancre. The poison is carried by the lymphatics to one of the inguinal glands in the groin. The poison here acts on the organic nerves, producing all the effects of inflammation. The poison is again absorbed by the lymphatics and veins, gets into the circulation, and, in about two months, poisons the organic nerves of the tonsils. In two or three weeks afterwards, perhaps more, the patient will be attacked with symptoms of fever, and in a few days he will be seen covered over with blotches. If a medical man is not on his guard, he may pronounce the patient as having measles or small-pox (!) and only know the true nature of the case until the scaly, copper-colored spots meet his eye.

After the eruption has existed for some time, iritis is apt to seize on one of the eyes. The patient will be attacked with all the constitutional symptoms denoting inflammation, such as a rigor followed by heat of skin, dry tongue, loss of appetite, thirst, quick pulse, and headache. He will complain that he has got cold in his eye; that he cannot look at the light without the tears flowing, and that it gives him pain to do so; he will request that something may be done for a severe pain in the temple, which terribly harasses him during the night. On looking into the eye, slight haziness of the cornea, a white ring round the cornea, as well as a dark-colored red zone immedi. ately external to the former, gradually shaded off, will be observed. The pupil will present an angular appearance; the iris will be changed in color, and appear puckered from the deposition of tubercles of lymph, of a yellowish-brown color. As the case goes on, lymph will be copiously deposited, filling the anterior as well as posterior chambers; occlusion of the pupil takes place; the cornea being pressed on it, loses its brilliancy, and becomes opaque.

I have now endeavored to describe the constitutional symptoms and appearances of acute syphilitic iritis.

Again—a patient may have passed through the various stages of secondary syphilis, and go on apparently well for a year, or much longer, when he will be tormented with severe pains at night in his collar-bones, elbow, and shin-bones; on examination, the periosteum will be found thickened. It may happen, too, at a later period, that he will direct attention to one of his testicles, when he will be found laboring under hydro-sarcocele. Here, also, iritis, generally of a chronic character, will present itself. The thickening of the periosteum, the enlargement of the testicle, with the effusion of serum, as well as the deposition of lymph in the iris, are attributable to the morbid action of the organic nerves in the periosteum, the testicle, and the iris.

To show that the syphilitic poison may be communicated by

the semen secreted under the influence of the organic nerves of the testicle, whilst imbued with the syphilitic virus, to a woman during coition, by the poisoning of the organic nerves of the vagina and cervix uteri, I will briefly allude to two cases, where the semen actually communicated syphilis to sound and healthy young women. And here I will add, by way of parenthesis, they were not French ladies, and were as respectable as the cases reported by Professors Parker and Porter. In fact, they were persons on whom the slightest shadow of suspicion could not rest.

Mr. C. was under my care for *syphilitica psoriasis guttata.* I ordered him compound decoction of sarsaparilla with the bichloride of mercury. He got well, with the exception of two small condylomata near the verge of the anus. He now said he should get married in a few days. I told him I did not think it safe to do so, and advised him to consult some other person. Accordingly, he took the opinion of one of the ablest and best surgeons in the metropolis, who gave him a written note, stating there was no apprehension to be entertained about his entering into the marriage contract. It is to be noticed there was not the slightest appearance of a sore on the penis. In about nine weeks after the marriage, the lady, with whom I was well acquainted, applied to me with an ulcerated throat and the copper-colored scaly eruption. Knowing how matters stood, I made no allusion to the genital organs, fearing I might arouse suspicion.

Mr. I., a young man, applied to me, having ulcers on his legs and other parts of the body. He told me he was ten weeks married; that he had the "bad disorder" before he was married, but that "the doctor" said he was cured, and might get married.

I examined his virile organ, and found it all right. In consequence of stating his fear that his wife got the disease from him, I requested to see her, and found her with an ulcer in the throat, and the copper-colored eruption. I inquired if she had any ulcers on the genitals, and on her replying she had not, I made no further examination.

Before speaking of the treatment of iritis, it is necessary to state, there are other forms of ulcers on the genitals, followed by secondary symptoms, including iritis.

Mr. Carmichael describes a superficial ulcer without induration, but with elevated edges. A similar ulcer, destitute not only of induration, but of elevated edges.

In the second class, Mr. Carmichael included the phagedenic and sloughing ulcers. He found that the papular eruption, with superficial ulceration of the throat and iritis, followed in one class; whilst phagedenic ulceration of the throat, pustular eruption, and rupia, followed in the other, as well as iritis.

It does not come within my province to enter into a discussion relative to the original views promulgated by Mr. Carmichael. I have therefore simply to observe that, during the years I was attending as a pupil, and acting as a dresser in the Richmond Surgical Hospital, Dublin, I had ample opportunities of witnessing the truths enunciated by Mr. Carmichael, in reference to syphilis, in his work published in 1814.

With respect to the treatment of syphilitic iritis. In the event of the patient being a strong, plethoric countryman, I would have recourse to venesection, quick mercurialization, the application of belladonna to the upper and lower eyelids, and follow up the treatment with iodide of potassium and the compound decoction of sarsaparilla, and keeping the patient confined to a dark room.

It may be asked how the blood-letting acts? It is well known the abstraction of blood is attended with a quiescent state of the vascular system. The stimulus is withdrawn from the organic nervous system—hence the tranquillity which supervenes.

Here it is right to observe, venesection is only justifiable in any case of inflammation, of any organ, immediately after the rigor has taken place, announcing that an attack has been made on the whole organic nervous system—indicated by the pale countenance, the cold surface of the body, the gnashing of the teeth, the feeble pulse, and extreme prostration.

When reaction is fully established, it is rendered manifest by the flushed countenance, the heat of the skin, the dry tongue, the thirst, the bounding pulse, the throbbing of the heart, the headache, and pain in some particular spot.

6

Under such circumstances, it will be perceived a violent struggle is going on between the organic nervous system and the blood. The organic nervous system during the rigor was in a depressed condition; and on its recovery from the shock, becomes violently excited—over-heats the blood, which in its turn over-stimulates the organic nervous system.

The pulse is the great index to inflammation. The phenomena concomitant on the former, when the latter exists, are now susceptible of lucid explanation on clear physiological principles. The blood stimulates the organic nerves in the internal coats of the arteries, causing the arterial tubes to contract firmly, which is moving with increased velocity and force. Hence the firm, and, at the same time, bounding pulse. Here, in fact, are two antagonizing forces: the artery pressing from without, the blood expanding from within.

Here the abstraction of blood restores the natural relations subsisting between the blood and the organic nervous system, and should be carried sufficiently far to indicate complete nervous tranquillity; and thus anticipate nature in the formation of a process to remove the offensive products from the system.

Under such circumstances, the old physicians grappled scientifically and vigorously with inflammation by copious bloodletting at the onset, and cut short the disease.

With what amazement would one of the old school contemplate a fashionable young physician of the present day, ordering Sherry wine or French brandy in such a case!!!

Here it may not be uninteresting to inquire, How do marsh miasmata produce intermittent fever?

How do the emanations from human effluvia induce typhus fever?

The answer is, the poison being an immaterial one, is incorporated with the atmosphere, is inspired, enters the blood with the oxygen, which is derived from the air; is again given off with the oxygen, by the capillary arteries, to the organic nerves. After a certain interval, the poisonous influence is fully brought to bear on the organic nerves, which is rendered apparent by the rigor and constitutional disturbance succeeding.

The same explanation is true of measles, scarlatina, and

small-pox; in the latter, the papilla, the vesicle, and pustule are produced in the same manner as chancre—by the poisoning of the organic nerves of the skin.

Mr. Hunter considered inflammation a healthy process. The formation of an abscess into which pus is secreted, shows the organic nervous system has overcome the shock, and has contrived a new organ for the purpose of secreting the pus from the blood—the pus contained in the cyst of an abscess may well be compared to the bile in the gall-bladder.

Here it will be observed, as soon as the organic nervous system has commenced the work of recuperation by the effusion of lymph, serum, or pus, that bloodletting should not be practiced —as it would interfere with the salutary process of nature. Here, too, it is that stimulants are required to prop up the exhaustion of the organic nervous system, which has been subjected, in the first instance, to a violent shock, next to over-stimulation, and lastly to weakness consequent on the latter, by the secretion of serum or pus.

I cannot avoid remarking, although it is not connected with the subject under discussion, that veratrum viride, aconite, belladonna, and tobacco arrest inflammation, by their sedative influence on the organic nervous system. I have not included opium, inasmuch as it appears to stimulate the organic nervous system, as is witnessed in the contraction of the pupil. Its operation is confined more to the animal nervous system, inducing sleep, and allaying pain.

When an old debauchee, who has been addicted all his life to habits of intemperance—who has often indulged at the shrine of impure venery—whose constitution has been saturated with mercury—who presents the characters of scrofula, or exhibits a tendency to phthisis—it would be adding fuel to the fire to give him mercury. I would therefore order him turpentine, as recommended by Mr. Hugh Carmichael, of Dublin, and apply belladonna round the eye. If the turpentine did not succeed, I would give him iodide of potassium with decoction of sarsaparilla, as recommended by Mr. Lawrence.

Professor Bennet states he has cured every form of iritis, including syphilitic, without mercury. I think the internal administration, joined with the external application of belladonna, after depletion by bloodletting, (if the patient should be vigorous,) and purgation by brisk cathartics, ought to cure iritis.

It is useless to apologize for the manner in which this paper is written. I will only add, had I more time at command, I could have done the subject more justice, and written it in a better style.

NOTES.

[1] A small organic nerve passes from the superior cervical ganglion to the iris, so that it is true the iris receives no animal nerves.

[2] In a case of chancre, or in a case where vaccination is practiced, the poison is applied to a particular part, so that the organic nerves of that particular part only are contaminated. In the case of small-pox, the poison accompanies the oxygen into the blood, which traverses the whole body, to give off the oxygen to the organic nerves; hence it follows as a sequence, that hundreds of pustules present themselves. In the one case, the organic nerves are *locally*, in the other they are *generally*, implicated. The poison in the former case is applied from without; in the latter from within.

This subject is most important, as well as attractive. The mode in which erysipelas, puerperal fever, yellow fever, and phthisis are propagated, can now be easily studied and understood. A distinct immaterial poison is the primary cause in each case.

The Nervous Centres of Animal and Organic Life.

It appears to be unquestionably true that in one class of animals, known amongst naturalists as the *Invertebrata*, there is only a single nervous system.

In the second class, called the *Vertebrata*, there is a *double* nervous system.

It is evident that the nervous system, which is common to both, must be the *more important*.

It would seem, therefore, that the second nervous system, which is only found in the higher classes of animals—(the *Vertebrata*)—was added to furnish a habitation for an *intelligent, immaterial* agent, capable of providing for the wants and sphere of life appertaining to their more perfect animal organization.

To understand this explanation, it becomes indispensably necessary to be thoroughly conversant with the distribution of the nervous system in the lowest classes, with a view of comparing them with the nervous systems of the highest classes of animals; and thus be enabled to note what parts or centres of the nervous system are present in each.

A concise description of the nervous system of man may not be misplaced in the first instance; and most probably will afford facilities towards comprehending the subjects to be subsequently brought under notice.

The cerebrum, cerebellum, pons varolii, medulla oblongata, spinal cord, with the cerebro-spinal nerves, form the animal nervous system.

The superior central, (Pineal gland,) the cerebral, (pituitary gland,) the otic, the lenticular, the spheno-palatine, the submaxillary, the three cervical, the thoracic, the lumbar, the sacral, the cardiac, the splanchnic nerves, the semilunar, the diaphragmatic, the stomachic, the hepatic, the splenic, the pancreatic, the nephritic, the mesenteric, the lacteal, the spermatic or uterine ganglia, with their branches and plexi, constitute the organic nervous system.

It is proper I should state that *I named* the ganglia, located at the anterior border of the semilunar ganglia, after the organs they were instituted to preside over.

Mr. Harrison, in his Practical Anatomy, observes that almost all the nerves of the cerebro-spinal system communicate with the organic nerves.

I will now proceed to demonstrate that one nervous system is sufficient to discharge all the functions appertaining to life in the Invertebrata.

In some of the lowest classes of animals, very great difficulty has been experienced by anatomists in discovering a nervous system of any kind.

Trebly, Goede, and Carus failed to discover any trace of a nervous system in the Acalepha.

Dr. Grant describes a nervous system which he found in the Boroe Pileus, consisting of a double circular nervous filament, situated round the oval extremity of the body, which sends off minute filaments in each of the spaces between the longitudinal bands of ciliæ.

These eight points, from which the longitudinal filaments come off, present ganglionic enlargements.

Spix, a German anatomist, describes a nervous system in the Actinia, which may be considered an isolated polypus, having no calcareous skeleton, and fixing itself in the rocks by its fleshy base, consisting of minute filaments, with minute ganglia surrounding the fleshy base, from which were were given off nerves to the different parts.

Amongst the Echinodermata, Tiedeman describes in a small species of this genus a nervous system, consisting of a circular cord around the mouth, from which proceeded a filament along each ray, having at its origin a minute ganglionic enlargement. The nervous ring rested upon the extreme edge of the central aperture, in the calcareous frame-work of the body, and the filaments rested on the inferior surface of the rays, concealed by and at the base of tubular feet and suckers.

Two other filaments, much shorter than those just described, are given off from each of these ganglionic enlargements, to be distributed to the stomach and other viscera.

This animal possesses considerable muscular power.

In the Ascidia Mammillata, belonging to the Mollusca Tunicata, Cuvier describes and figures the nervous system as consisting of a single oblong ganglion, situated near the anus of the animal, and between *that* and the bronchial orifice. From the ganglion branches are given off; some of which, passing to the œsophagus, encompass it in the form of a ring.

This animal is surrounded by a muscular sac, which, by its contraction, can compress and empty its general cavity. This receives some muscular filaments.

The solitary ganglion of the Ascidia seems to regulate the action of the orifices of ingestion and egestion, and of its enveloping sac, on which depends the slight locomotive action of the free species.

Dr. Anderson says, in the Conchifera the nervous system is adapted for the functions these animals have to perform, which are: ingestion of the food, respiration, and locomotion.

These nervous centres, or ganglia, are consequently placed in immediate relation to the organs destined to those functions.

The œsophageal or labial ganglion are the most important. They are two in number, situated more or less near the mouth, and are united by a transverse band, which arches over it.

From these ganglia nerves are given off to the mouth and tentacles, and to the anterior parts of the viscera.

Each ganglion has a branch of communication to the pedal ganglion and to the bronchial ganglion.

In the Distoma Hepaticum, belonging to the Entozoa, Bogannus describes a nervous system, consisting of a nervous collar or ring, with two lateral ganglia entwining the œsophagus; and two nerves, which are distributed to the posterior part of the body.

Otto describes the nervous system of the Strongylus Gigas as consisting of median nervous filaments, with closely approximated ganglia.

The Cirrhopoda have abdominal cords, with ganglia developed on them—and there is a nervous collar around the œsophagus.

The Annelida have a varied number of ganglia. united by double longitudinal fissures.

In the Crustacea, the common Talitrus has a regular series of ganglia developed, at an equal distance from each other.

In the Myriapoda, the Scolopendra Morsitans has a nervous system consisting of twenty-one double ganglia, situated on the ventral surface of the body, connected by intervening double longitudinal cords. From each ganglion are given off lateral nerves, to supply the neighboring muscles, viscera, and feet. These ganglia are nearly all of an equal size, excepting the first, which is the largest, and from which are given off additional nerves to supply the maxilla.

Mr. Owens says each joint of the Articulata corresponds to a division of the nervous system.

In the Gastropoda, the common snail has two nervous centres: one placed *above* the œsophagus, the other below it—both connected by two cords, embracing the œsophageal tube. The upper ganglion supplies nerves to the muscle of the mouth, as well as the skin in its vicinity.

It likewise furnishes the nerves of touch, and of vision, besides those distributed to the generative organs. And from the sub-œsophageal ganglion, which fully equals the brain in size, arise those nerves which supply the muscles of the body and viscera.

The nervous centres obey the movements of the mass of the mouth, with which they are intimately connected. They are pulled backward and forward by the muscles, serving for the *protrusion* and *retraction* of the oral apparatus, and are thus constantly changing their relations with the surrounding parts.

In the snail it would seem that the great mass of the nervous collar which embraces the œsophagus will, in some instances, permit the mass of the mouth to pass entirely through it, so that sometimes the brain rests on the œsophagus, and other times it is placed on the inverted lips.

In the Nudibranchiate, the nervous centres exist in the most concentrated form, and indeed it is doubtful whether there are any other ganglia, excepting the large supra-œsophageal ganglion.

In the Tritonia there are four tubercles placed across the commencement of the œsophagus, the nervous collar being completed by a simple cord.

All the nerves which supply the skin, the muscular integument, the tentacles, the eye, and the muscles of the mouth, arise from the tubercles. And anatomists have not hitherto detected any other source of supply. For these particulars I am indebted to Dr. T. Rymer Jones.

It would be superfluous to continue giving examples of animals having only a single nervous system.

The next matter, therefore, should be to determine what division of the nervous system in the vertebrata is identical with, or discharges the same functions as, the nervous system of the Invertebrata.

All animals require for the continuance of life the ingestion of food, or nutriment, in the stomach. Therefore, in all animals there must be provision made for the function of deglutition, and such is found to exist in all animals.

The nervous rings surrounding the mouths of the lowest classes of animals, such as the boroe pileus, the star-fish, the œsophageal ganglia found in the mollusca, preside over the function of deglutition in the invertebrata.

The location, the position, the distribution of the nerves, point out the spheno-palatine ganglion, as presiding over the function of deglutition in the Vertebrata.

It will be recollected that this is one of the organic ganglia. I will now quote Mr. Harrison's description of the spheno-palatine ganglion. "It is a small, triangular, reddish substance. It is imbedded in fat, surrounded by branches of the internal maxillary artery, and is situated on the external side of the nasal plate of the palate-bone, which separates it from the cavity of the nose, behind the tuberosity of the superior maxillary bone, and in front of the pterygoid processes. Three sets of branches pass from the ganglion: an inferior, internal, and a posterior.

First, the inferior, or palatine nerves, descend in the bony canal of that name; send through the canal some small twigs, to the spongy bones, and near the palatine separate into three filaments; an anterior, middle, and posterior. The anterior is the largest, and passes forward in a groove within the alveoli, and above the mucous membrane, supplying the latter, the bone and teeth, and finally enters the foramen incisivum by a very fine filament, which communicates with the nerves in the septum narum.

The middle and posterior filaments of the palatine nerves are distributed to the amygdalæ, the soft palate, and uvula. The posterior usually decends through the osseous canal of the pterygoid portion of the palate-bone. The internal branch of the spheno-palatine nerve is very short, passes through the spheno-palatine hole to the upper and back part of the nose, and divides into five or six branches. The most important of these pass immediately into the mucous membrane covering the superior and middle spongy bones; one branch, called the naso-palatine nerve of Cotunius, passes beneath the sphenoidal sinus, across the root of the nose, and decends obliquely forward, along the septum nasi, as far as the foramen incisivum, where it communicates with the interior palatine branches, and where some anatomists describe a small ganglion (naso-palatine) to exist. This, however, in the human subject, can seldom be distinguished from the surrounding fat and vessels.

The third, or posterior branch of the ganglion, is the vidian, or superior petrosal nerve; this passes backward through the Vidian canal, above the pterygoid plate, and sends some small filaments into the sphenoidal sinuses; it there perforates the cartilaginous substance that closes the foramen lacerum anterius, enters the cranium, and divides into branches; an inferior and posterior. The inferior, or carotid branch, enters the cavernous sinus, and joins the plexus formed around the artery, by the ascending branches of the superior cervical ganglion of the sympathetic. The superior, or petrosal branch, runs backward and outward, beneath the dura mater and Casserian ganglion, in a groove on the petrous bone, enters the hiatus Fallopii in the bone, and becomes attached to the portio dura nerve—the part of function being marked by a small gangliform expansion. The Vidian nerve accompanies the portio dura as far as the back part of the tympanum; it then leaves it, receives the name of chorda tympani, and enters the tympanum a little below the pyramid; invested by mucous membrane, it now proceeds forward, between the long leg of the incus and the handle of the malleus—to the latter it is firmly connected; it then escapes by a canal, which appears near the internal extremity of the glenoid fissure; it next runs downward, inward, and forward, joins the gustatory nerve, and continues attached to it as far as the submaxillary gland; it now leaves the gustatory nerve, and unites with filaments from it in the submaxillary ganglion, which is situated near the posterior edge of the submaxillary gland, and from which a number of filaments proceed; these form a plexus, which supply the gland."

It requires very little reflection to understand the part the intelligent agent, situated in the spheno-palatine ganglion, has to enact in the process of the selection of food, mastication, the salivary secretion, and deglutition, when its connections are remembered.

It may be supposed that sufficient proof has not been given to sustain the declaration that the œsophageal ganglion in the invertebrata, and the spheno-palatine in the vertebrata, are identical, as to their functions in the animal economy.

I beg, therefore, to direct attention to comparative anatomy.

In the Ruminantia, the spheno-palatine ganglia are very large;—they are double the size in a sheep, when compared with a carnivorous animal of similar dimensions. I presume it is unnecessary to specify (the mode of mastication in these animals is quite different from the carnivorous) or to discuss the reasons why the spheno-palatine should be so large in the former, as reflection will at once suggest the answer.

Again, it will be recollected that a snail, having a supra-œsophageal ganglion, has the power of regurgitation.

In fact, it can swallow its oral apparatus, and again regurgitate it. In this respect it resembles the Ruminantia. It is a remarkable fact that several of the Gasteropoda, to which the snail belongs, have stomachs similar to the Ruminantia.

I have next to observe that the circulation is common to the invertebrata and vertebrata. As the former have only a single nervous system, it is evident that the nervous centres which preside over similar functions in the vertebrata, must be of the same character as that of the Invertebrata.

In the human subject, I think I will be able to show satisfactorily that the heart and arteries are under the superintendence of the intelligent agent situated in the organic nervous system.

"The cardiac ganglion," says Mr. Quain, "lies beneath the arch of the aorta, and the bifurcation of the trachea, in close contact with the former, extending from the division of the pulmonary artery to the origin of the brachio-cephalic. This may be considered as the common point of union of the cardiac nerves that issue from the cervical ganglion, and the immediate source from which the different nerves proceed to supply the heart."

Scarpa thus describes the cardiac ganglion : " Anastamosis illa, valde insignis, quæ inter utriusque lateris cardiacorum nervorum truncos, sub aortâ curvatura, paulo supra cor conficitur."

Mr. Harrison remarks: "The size and structure of the cardiac ganglion are very variable. Instead of a single distinct ganglion, it often appears as a congeries of small ganglia entangled in the plexus of the uniting nerves."

Mr. Harrison also observes, "that the roots of the large vessels and the structure of the heart are supplied by branches from the great cardiac ganglion and plexus."

In my opinion, the congeries of ganglia should be called the *right auricular;* the *right ventricular;* the *left auricular;* the *left ventricular;* the *aortic* and *pulmonary* ganglia.

These ganglia being all united, form the congeries that Mr. Harrison speaks of, and send off a plexus of nerves to the organs just designated, so that each can discharge its functions independently, and at the same time, harmoniously. Thus it is that the four apartments of the heart are located in one muscle; and thus, too, do the cardiac nerves of the right and left side terminate in ganglia, united together, apparently forming one body, like the heart, but destined to preside over the left and right sides of the heart.

The mechanism of the heart, as well as the arrangement of the ganglia, challenge and demand the most profound thought of the anatomist and physiologist, as being the most exquisite and beautiful piece of workmanship in the human body, and the tabernacle in which are placed thrones for rulers, possessed of extraordinary and unerring wisdom.

I have now stated sufficient facts to enable any person to perceive that the heart, the great central muscle of the circulation, receives its supply of nerves from the organic nervous system; and consequently, that the nerves connected with the circulation in the Invertebrata must be of the same kind; and such, in truth, is the fact, the *invertebrata* having only one nervous system.

Respiration is common to both classes of animals. The blood is aerated in each by the pulmonary nerves, derived from the anterior and posterior plexus.

I am satisfied the pulmonary nerves have the power of decomposing the air or water, as the case may be; allowing the oxygen to pass into the blood, and the carbon to be set free. The oxygen is necessary for the maintenance of life in the organic nervous system.

It must strike every person that, if oxygen were to pass in, and carbon to pass out at the same moment, carbonic acid would be formed; thus no oxygen could get into the blood.

It will be recollected that the TORPEDO has the faculty of giving electric shocks. As electricity will decompose water, it therefore follows that the *torpedo* has the power of decomposing water into its separate elements.

As the electrical organization is connected with nerves in the fish alluded to, so in like manner is the decomposition of air in the lungs accomplished by nervous influence.

Dr. John Davy has obtained decisive evidence of chemical agency being excited by animal electricity. He passed the discharge from a *torpedo* through a solution of nitrate of silver, common salt, superacetate of lead, and found that all were decomposed.

I hope I will not be deemed as digressing from my subject in making the following remarks.

The *torpedo*, it is known, must be in a state of vigor and activity to give off the electric fluid with force. A man must be healthy and strong before he can have proper semen discharged.

The *torpedo* must be stimulated; so, too, must the man. When the semen is emitted, a distinct shock is communicated to the whole system. Therefore, the production of the semen is *isocronous* with the shock.

The power which gives the shock, on the one side, gives the impulse to the semen on the other, which is promptly ejected.

As life is situated in the organic nervous system, nothing else can be imparted to the semen but *life itself*.

It therefore follows, that the semen contains the vital agent. I will now relate an anecdote, which, although ludicrous, will throw considerable light on this interesting subject.

A cook lived in a family in Ireland, in the neighborhood of the place in which I was born, where a fool, of a very low degree of intelligence, was daily in the habit of visiting. Being of an amatory disposition, the cook indulged the appetite of the fool with the good things of her department, and eventually induced him to have sexual intercourse with her. All went on well until the semen was being discharged, when the fool exclaimed in *Irish*, which is a very expressive language, " Murder! murder! the life is leaving me!"

This story is literally true, and explains forcibly the shock given to the nervous system, as well as the impression conveyed that something more than the semen was evolved.

The semen being the result of a process, where the whole organic nervous system is engaged, and as the influence of the whole is concentrated on one particular organ, it so happens that, cotemporaneously with the shock, a miniature representation of the exterior of the body, as well as a true delineation of its internal organization, is daguerreotyped on the spermatic ganglion, which is instantly reflected on the semen, and propelled with force, in many cases, to the nidus prepared for its reception. Hence family likenesses between parent and offspring can be accounted for.

I repudiate the superlatively Liliputian idea entertained relative to the spermatozoa, as contrary to reason and common sense. The enthusiasts that state that there are myriads of living animals in a drop of semen, each capable of being formed into a man, must hold the opinion *too*, that the one left to become fully developed has been guilty of the most unpardonable fratricide, in annihilating his innumerable brethren, who were placed in a similar position with himself, as to rights of independent existence.

Lallemand has adduced several instances of a fœtus seven months old, in which all the functions of life were carried on, and the organization completed, although the brain and spinal cord were absent. It will be perceived that a fœtus of this kind precisely resembles a MOLLUSK.

I am indebted to Dr. Batchelder, a highly distinguished Fellow of the Academy, for kindly suggesting the proof that even in the human subject, the cerebro-spinal system is not necessary for the continuance of life.

It now becomes necessary to disprove the theory that respiration, circulation, and digestion depend on the operation of the par vagi.

The vagi may be divided without stopping respiration, circulation, or digestion. Longet operated on dogs, some of which lived to the fifth day. Dupuy found that horses lived up to the seventh day. De Blainville, that pigeons lived to the seventh day.

These experiments alone are conclusive that the functions specified do not depend on the vagi.

Several distinguished physiologists failed in exciting the muscular action of the heart by irritation of the vagi.

Longet mentions that he failed in influencing the rhythm of the heart by the application of galvanism to the vagi of dogs, rabbits, and sheep; but very frequently succeeded by scraping the cervical cardiac branches of the vagus. It is to be remarked that Longet did not draw any distinction between the animal and organic nerves; he forgot, or did not appreciate the fact, that the cardiac nerves were derived from the cervical ganglia, and consequently of precisely the same character as the nerves distributed to the heart.

This experiment proves conclusively, and positively, that the organic nerves are only influenced by the action of the organic nerves.

I cannot use language sufficiently emphatic to impress on physicians and surgeons to pause and reflect on these experiments on the vagi and cardiac nerves.

Legallois has proved by numerous experiments that an animal will continue to breathe after the division of both vagi in the neck, if care be taken to secure the ingress and egress of air to and from the lungs.

Mr. Reid observes, that if the vagi be injured above the origin of the recurrent laryngeals, none of the muscles attached to the aretenoid cartilages can any longer act in unison with the muscles of respiration—all these movements cease, and the superior aperture of the larynx can no longer be dilated during inspiration. Let me here state that the organic nerves which supply the muscles of the larynx are derived from the superior cervical ganglion; and that the pulmonary plexus is partly formed by filaments from the long cardiac nerve derived from the superior cervical ganglion; and that division of the vagi, as above stated, destroys the unity of action between the organic nerves in the larynx and pulmonary plexus.

An animal nerve is capable of holding up communication between one organic nerve and another of the same kind.

Mr. Reid says, "Although respiration were much diminished by the removal of the cerebrum and cerebellum, and then dividing the vagi, they continued living for a longer or shorter time."

Volkman, Flouren, and Longet confirm these observations by experiments.

Mr. Reid confirms the experiments of Dupuytren, that no morbid change could be discovered in the lungs of dogs, on the side on which the vagus had been tied, in six months after the operation.

Messrs. Mayo and Müller failed in exciting muscular contraction in the stomach by irritating the trunk of the vagi.

Bichat, Tiedeman, Gmellen, Longet, Breschat, Milne, Edwards, inferred that the muscular movements can be excited in the stomach of a living animal, by galvanizing the lower end of the vagi in the neck, from its effects upon the digestive process.

These gentlemen seem to have forgotten that the branches of the par vagi inosculate with branches of the stomachic plexus in the stomach, and that the secretion of the gastric juice depends on the operation of the latter nerves.

Majendie observed that these muscular movements of the stomach continued after the section of the vagi. Mr. Reid confirmed Majendie's remark by experiments on a dog; where, after cutting the vagi, and on the dog recovering, he found that the stomach could still propel the chyme onward towards the duodenum.

Messrs. Reid and Longet found that dogs, whose vagi had been divided, experienced sensations of hunger, if they survived a certain number of days.

Leuret and Lassaigne detail the result of an experiment on a horse, where the process of digestion went on after the division of the vagi, with loss of substance.

Arneman tied the vagi of a dog, and as the animal lived until the 165th day after the operation, it was killed.

Sedillot, Chaument, and Mr. Reid arrived at similar results from experiments on dogs—that the digestive process was carried on after the division of the vagi. What stronger proofs, what more forcible arguments, what clearer demonstration could be adduced than the experiments just detailed to prove that respiration, circulation, and digestion can be carried on without the assistance of the vagi?

That the nervous system in the Invertebrata is of the same character as the organic nervous system in the vertebrata, can be demonstrated by direct experiment.

Every joint of the class of animals known as the Articulata possesses a distinct nervous system, capable of carrying on all the functions appertaining to its individual capacity. Hence it is that a lobster may be partitioned into several parts, and each part be still living. The common earth-worm may be divided into parts, and each will be capable of forming a perfect animal. Numerous instances of this kind might be cited, were it necessary so to do.

As it may be supposed such results could not be obtained by section of the Vertebrata, I have to repeat what I stated in a former paper, that the brain of a sheep may be destroyed; that the head may be severed from the neck at the articulation between the atlas and condyles of the occipital bone; that the head will give evidence of life, by opening and closing the mouth;

7

that the body will give vigorous manifestations of life for minutes.

This experiment is conclusive that the cerebro-spinal system is not engaged in the vital functions, and further proves the identity of the ganglionic nervous system of the sheep with that of the Gasteropoda, to which class the snail belongs, and whose head can be cut off almost with impunity; the Articulata, to which the crab and lobster belong; the Annelida, to which the earth-worm belongs.

Having now, I hope, clearly shown that the nervous system in the Invertebrata is the counterpart, and fully and truly represents the organic nervous system in the vertebrata, I will next proceed to show the functions of the cerebro-spinal system.

That an intelligent, immaterial agent resides, or has its habitation in the brain, is momentarily demonstrated; and it scarcely requires to point out that the nerves of sense are connected with it. To elucidate the fact that it is the seat of the mind—it is by the full development of this animal nervous system that man shows his superiority over all other animals, in wisdom and reason. The mind becomes conscious, by the organs of sense, of the nature of all bodies external to it. The mind reflects, judges, and wills what course should be pursued by the body, under certain circumstances; and has under its control a chief messenger or agent, known as the spinal cord—which latter is amply supplied with auxiliaries, in the shape of spinal nerves, which proceed from it to all parts of the body—fully carrying into execution the commands with which they are charged by the mind, as well as carrying back to it any intelligence which it should be made cognizant of.

The animal, intelligent, immaterial agent, inhabiting the brain, requires, and is susceptible of receiving knowledge and education, and has provision made for its reception, by an expansion of the cerebral matter.

Without entering into a discussion relative to Phrenology, it may be stated that, in proportion to the development of the cerebrum, the mental intelligence is of a very high or low standard.

It is a well-known fact that idiots have thick skulls—that

the cerebrum is not developed to any considerable extent—that its place is sometimes occupied by a cyst. In some of the Hottentot tribe, the cerebrum is not of the same size or dimensions as the European. The deficiency in intellect, as well as moral endowments, are too well understood to require any force of argument to substantiate. Here, it must be admitted, a physical impossibility is placed in the way of the race, that these people form a portion of, to cultivate the arts, the sciences, industry and morality, that characterize and distinguish the white population.

Having, I trust, proved the correctness of Bichat's theory, with respect to there being two nervous systems, an animal and vital, I will endeavor to point out their connection and mode of action.

What can be more magnificently grand or transcendently sublime, than the examination and contemplation of the scientific and ingenious arrangement by which two intelligent immaterial agents are made the occupants of two distinct, material substances, perfectly developed in their organization, and merely connected together by links or bands of communication—and thus ordained to discharge their functions, in mutual unison and usefulness?

How provident the order which commands the one to repose, thus insuring a quiescent state of the entire members of the body, which are subjected to physical exercise by the mandate of the will!

How wise the precaution that precludes the possibility of the other postponing its operation!—any cessation being incompatible with the continuance of life!

What a careful conservative principle is manifested in the design, which enables the immaterial agent, resident in the organic ganglia, being placed in such a position as to be enabled to regulate and assist its fellow, situated in the animal nervous system, whose duty consists in providing for communication with the external world, thus acting ostensibly in a capacity subservient to life, or its requirements!

How marvelous the sagacity and premonition which exclude and prohibit any direct communications taking place between an organ essentially associated with life and an instrument of the will!

If the heart, for instance, depended on, or was under the control of the will, what enormous mortality would be the result! Good care, however, is taken that no animal nerve should visit the heart; thus excluding the animal, immaterial agent, and leaving the vital, immaterial agent supreme ruler of the action of the heart; and thus it will be perceived that the organ on which life depends is placed under the guardianship of life itself, and is perfectly independent of the will.

I am not indulging in figurative language, but simply stating facts, which have already been proved by direct experiment.

It would be superfluous to enter into details of the reasons, or the results, of the vivisections set forth in my former papers, to prove the Pineal gland is the great central ganglion, or president of the organic nervous system. On the present occasion, suffice it to say, that *it is;* and that, when the operations of the mind are active in the brain, there is a reciprocal communication taking place between the immaterial agent, located in the brain, and the other immaterial agent resident in the ganglia. Whatever troubles the immaterial agent in the brain, likewise harasses the immaterial agent in the organic nervous system.

Let any man place his hand over his heart and deny this proposition if he can. Who has not experienced anguish, depression, and painful sensations about the heart under certain bereavements? Why are not similar feelings experienced on the right side? No one supposes, I believe, that the muscular substance of the heart is the source from which the grievance emanates. Who can now be at a loss in accounting for the source, when he can point with the tip of his finger to the site of the great cardiac ganglion?

If the mind, located in the brain, can communicate with the cardiac ganglion through its connection with it by the par vagum, *à priori*, the brain can communicate with the central ganglion through the pedunculi of the brain attached to it.

The central ganglion may be pronounced as a ganglion *sui generis;* the same observation is true of the cardiac ganglion.

Every practical physician and surgeon has ample opportunities afforded him, almost daily, of witnessing, what may be de-
nominated an experiment, on a living subject, which should convince him of the relations subsisting between the brain and the
ganglion; and which should have long since attracted the attention of some astute and thinking physician.

In the first stage of meningitis, there is contraction of the
pupil—in the last stage there is dilatation of the pupil. At *first*
there is irritation of the brain; and *lastly*, there is effusion into
the ventricles of the brain. The brain is connected by the par
vagum to the cardiac ganglion, just as it is to the lenticular
ganglion, by an exceedingly small filament of the third pair;
but the brain cannot influence the pulsations or movements of
the heart, which derives its nerves from the cardiac plexus; and,
à priori, the brain cannot influence the iris, which receives its
nerves from the lenticular ganglion.

Irritation of the cardiac nerves, which are connected with the
branches of the par vagum, will influence the rhythm of the
heart—and why? Because they belong to the same class of
nerves as those which supply the heart. Irritation or compression of the central ganglion will affect the lenticular ganglion,
through the medium of the connecting branch of the third pair,
because the two ganglia belong to the same nervous system.

In further illustration of the laws which regulate the action
of the two nervous systems, the organ of hearing affords a beautiful example.

The otic acts as an assistant to the auditory nerve, and causes
the contraction or relaxation of the tensor tympani muscle; and,
of course, a similar condition of the tympanum, to meet the ne-
cessities or wants of the auditory nerve, in regard to communicating the different kinds of sound to the mind.

The lenticular ganglion, on the same principle, regulates the
admission of the rays of light through the pupil, to enable the
retina to execute its office efficiently in carrying the image of
the figure, impinged on it, to the mind.

The superior cervical ganglion sends a branch of communication to the lenticular ganglion; also a branch to the sixth pair,
as well as branches to the cervical nerves.

The reasons for these communications will be understood when it is recollected what occurs when a person looks behind his shoulder—the branches of the organic nerves, sent to the spinal nerves, cause the proper contraction of the muscles of the neck.

The branch sent to the sixth pair regulates the contraction of the abducens muscle—whilst the branch sent to the lenticular ganglion fixes the pupil in a proper axis to receive the rays of light—thus harmonious action of all the muscles is insured.

What a beautiful and admirable arrangement!

Again, this ganglion sends nerves to the muscles of the larynx, to the cardiac ganglion, which sends branches to the heart and lungs.

The mode in which the parts act will be perceived, on reflecting what occurs when a person is within the sphere of carbonic gas.

The arctenoid muscle at once contracts, closing the rima glottis—thus the action of the heart and respiration cease at the same moment—instant death is the consequence.

The spheno-palatine ganglion has already been described. Its functions consist in causing a secretion from the submaxillary gland, of saliva, to be incorporated with the food during the process of mastication; causing a secretion of mucus from the tonsils, for the lubrication of the bolus in its passage to the œsophagus; to prevent the food passing through the posterior nares, by throwing over them the curtain formed by the soft palate; it also prevents jarring of the teeth.

The branches of the ganglion act in concert with the fascial, as is exemplified when a person sees food he relishes; the saliva is secreted. It acts also with the gustatory branch of the fifth pair, in selecting the food which should be eaten, as well as that which should be rejected. Thus it is that animals are able to avoid poisonous herbs, a circumstance which is well marked in the sheep and goat.

The cardiac ganglion gives off the plexus which presides over the action of the heart, and holds communication with the central ganglion, through its connection with the par vagum.

The pulmonary nerve presides over the aeration of the blood;

and, by its connection with the par vagum, has the power of calling to its assistance the respiratory muscles, when any obstruction to the entrance of air into the bronchial tubes exists.

The idea that the construction of the muscular fibres of the intestines depends on irritability of the muscular fibre, as well as a similar condition of the uterus, by a like cause, will be found to be fallacious, when critically examined.

It is an ascertained fact that the iris is a circular muscle, and receives nerves from the lenticular ganglion. It is equally true that the intestines are composed of circular fibres, and receive nerves from the mesenteric plexus or ganglion.

The uterus is a hollow muscle, composed of circular, longitudinal, and diagonal fibres, supplied with nerves from the uterine ganglion.

If, therefore, the iris contracts under the influence of the nerves derived from the lenticular ganglion, it follows, as an irresistible deduction, that the intestines, as well as the uterus, being similarly circumstanced, with respect to the distribution of nerves, will contract on the same principle.

In the one case, there is ocular demonstration of the action of the nerves, as is seen in the eye.

In the others, there are palpable and unmistakable evidences brought under notice.

I am now arriving at a stage of my investigation of the greatest importance and all-absorbing interest. It relates to a phenomenon with which every physician and surgeon should be perfectly familiar; and one, too, which in my opinion is susceptible of clear elucidation, if not actual demonstration.

I allude to the circulation of the blood through the heart and arteries, under the agency of the immaterial agent situated in the organic nervous system. A correct knowledge of the nature of inflammation, as well as fevers and other diseases too numerous to mention, is allied with a full conception of this matter. It becomes one, therefore, of vast magnitude. Here I cannot avoid repetition:

" From the cardiac plexus (I quote Mr. Quain) three orders of filaments proceed: some pass backward, and join the pulmonary plexus; others turn forward, to gain the forepart of the aorta; but the descending branches, by far the most numerous, pass to the heart itself. In fine, they are dispersed in two sets, which take the course of the coronary arteries, and are thence termed. The anterior coronary plexus passes forward, between the aorta and pulmonary artery, and ramifies on the right ventricle and auricle. A great number of them being directed towards the right border of the heart, whence they communicate with the branches of the posterior, or coronary plexus. This plexus will be found to ramify on the inferior and posterior surface of the left ventricle and auricle.

" These nerves were at one time supposed to be confined to the arteries which they accompany, but the researches of Scarpa have shown that they pass away from the vessels in many places, and enter the muscular structure of the heart."

Mr. Harrison gives a more minute description of that plexus, showing the connection of the cardiac nerves with the convexity and concavity of the aorta.

It requires no arguments to prove that the heart, as well as the aorta, in the first and second stages of its course, is amply and largely supplied with organic nerves.

Mr. Harrison says, in speaking of the filaments given off by the inferior cervical ganglion, " that several also encircle the subclavian artery, and unite beneath it in the first thoracic, or dorsal ganglion." Again, he observes, " some extend along the subclavian, or axillary artery and its branches, and may be traced to a great distance, forming plexi in their tissue. A considerable fasciculus ascends along the vertebral artery, forming plexi around this vessel.

" Those of opposite sides unite with the basilar artery; they follow its branches, and communicate with analogous filaments from the carotid plexus."

Mr. Quain says, in speaking of the superior cervical ganglion, " that the ascending branches are two in number; they enter the foramen caroticum, and form around the artery a plexus, from which two filaments pass upward, to communicate with the sixth nerve in the cavernous sinus.

" One or two may also be traced along the carotid artery, as far as the minute ganglia, placed on the arteria communicans;

another terminates in the PITUITARY GLAND and INFUNDIBU-
LUM."

Here, I beg to remark, the PEDUNCULI of the PINEAL GLAND,
or CENTRAL GANGLION, can be traced in the brain of a sheep
down towards the INFUNDIBULUM, thus forming a complete nervous communication between the cervical, cerebral, and central
ganglia.

Here the intimate connection of the organic nerves with the
arteries is sufficiently manifest to the most superficial observer.

A plexus of nerves, derived from the thoracic ganglia, surround the thoracic aorta. The semilunar ganglia, at their anterior border, will be found connected with eight or ten smaller
ganglia, connected together by filaments. These ganglia, already
named, taken together, form what is called the solar plexus, and
from which radiates a plexus of nerves to encircle the different
arteries, which spring from the aorta, in their course to, and
entrance into, the organs after which they are named.

Mr. Quain remarks, "The aortic plexus is the direct communication of the solar; its branches form a complete network
upon the aorta, which can be traced along the ILIAC vessels."

It is well known, that when any obstacle is thrown in the
way of the free circulation of the blood, the *vis-à-tergo* is increased, as is witnessed in valvular disease of the aorta, causing
constriction of the orifice; when the left ventricle contracts
more forcibly to overcome the difficulty, its muscular development is increased, and hypertrophy of the left ventricle is said
to exist.

When a person is attacked with inflammation under a strong
fascia, the artery leading to it will be found to beat strongly
to surmount the impediment preventing the circulation of the
blood; that is to say, the compression of the vessels underneath
the fascia.

The heart receives its nerves from the cardiac ganglion.

The radial artery, assuming the inflammation to be under the
palmar fascia, receives its nerves from the plexus surrounding
the axillary artery. Similar causes, similar agents, similar
effects, characterize and assimilate the former to the latter case.

It is generally supposed that the heart contracts by irritability of the muscular fibre. In reply to this theory, it may be stated, the iris contracts and dilates. Passing from a dark to a well-illuminated room, will cause the iris to contract; whilst returning to the dark room will cause it to dilate. Hence alternate contraction and dilatation can be produced.

The heart, with its several cavities, may be considered, taken as a whole, as a hollow muscle, capable of contracting and dilating, as is witnessed in the case of the iris.

The *iris* receives the ciliary nerves from the lenticular ganglion; the *heart* receives its nerves from the cardiac ganglion.

The nerves received by both muscles are of the same character; and hence the phenomenon in both cases must be attributed to the same nervous influence.

If the muscular, or what is usually called the cellular, coat of the artery receive organic nerves, then its contraction and dilatation can be accounted for, precisely on the same principle as the heart, or iris, or intestinal tube.

In the first Vol. of the Transactions of the Physico-Medical Society, of New York, for 1817, will be found a very valuable and ably written paper, entitled "Reflections on the Pulsations in Epigastrio, with an Inquiry into its Causes," by Valentine Mott, in which the following passages occur:

"That a pulsatory motion in the epigastric region should occur, unaccompanied with disease of any of the surrounding organs, is a *curious* and *interesting* fact. It is one of the most *extraordinary* and inexplicable phenomena attendant upon *nervous irritation.*"

Again: "That nervous irritation should here be *concentrated*, and develop itself in the form of a *pulsation*, is no more *extraordinary* than the phenomenon of BLUSHING."

Further on: "A very strong and regular pulsation was felt in epigastrio. It was so great, that Morgagni says he never saw it exceeded—it was very visible externally. The dissection of this patient showed no vestige of disease, either of the heart, large vessels, or abdominal viscera."

In the *London Lancet*, published in 1833, there is a case reported, which was under the care of Dr. Watson, in the Middlesex Hospital, of a tumor in the epigastric region, which was mistaken by several practitioners, who declared it to be aneurism, and which subsided on the patient being well purged.

What stronger proofs could be adduced to prove the contractility and dilatation of the arteries?

What higher American or European authority could be cited to show that the contractility and dilatability of arteries depend on nervous influence, than the illustrious Professor Mott?

Who, possessed of the organs of vision, could contradict the conclusion, that the pulsation in epigastrio depends on nervous irritation, on seeing the aorta and cœliac axis completely surrounded by nerves derived from the solar plexus?

I cannot avoid mentioning, in proof of the nervous power exercised over the muscular fibres of the heart, that when the heart of a reptile is cut out of the chest, it will *bound about* for some time, just as the joint of one of the Articulata will move or change its position. As it may be said that this action of the heart of the reptile depends on *muscular irritability*, it is necessary to state, that if the animal is killed by *strong*, or concentrated prussic acid, there will be no such movement of the heart. The reason is obvious, because the vital immaterial agent, residing in the organic nervous system, has been destroyed, or ejected from its entire habitation.

I have to remark that the organic nerves are *extremely* long and *delicate*, in comparison with the animal nerves. A good example of this is presented in the cardiac nerve, which proceeds from the superior cervical ganglion.

Again it will be remembered, in the *Invertebrata* the nerves will be found accompanying the arteries.

It might not be deemed a long stretch of the imagination to assert that the plexus of nerves, which can be traced to such a very long distance, encircling the arteries, accompany them to their destination, which is unquestionably true, as I will presently demonstrate from comparative anatomy.

Mr. Swan, in his admirable description of the organic nervous system in the *boa constrictor*, says, "This plexiform structure varies in different parts, and becomes much greater about the beginning of the intestines, but it resembles that corresponding in the semilunar ganglia in the turtle. Near the kidney it assumes the form of a *nervous membrane*, or *retina;* before it is distributed on the *urinary* and *generative organs, branches* pass from the plexi with the *arteries to the different viscera."*

Could a clearer demonstration be given, that the organic nerves surround the arteries in the shape of a *retina ?*

I presume it is not necessary to insist on the difficulty of distinguishing a *nervous expansion* as fine as the *retina*, from the *cellular coat of the artery*, or even to dissect it off when recognized.

It follows from what has been now stated, that wherever there is a *nerve* there is an *artery*, and where there is an *artery* there is *blood*.

This inquiry admits of being gone into in more minute details. However, sufficient data, it is expected, have been presented to convince the most incredulous that such a precise adaptation of ingenious means to accomplish wise ends, could not have been the production of *chance;* as well as, that the supporters of the doctrine of *spontaneous generation*, together with the chemists, who attempt to explain the phenomena of life on philosophical or chemical principles, *are in error.*

It is manifest, indeed, that the former class of persons never reflected on the intricacies of the organization, and peculiar manner of animal formation; whilst chemists, in arriving at conclusions, forget they are only experimenting on the effects of a certain cause, which is called *life.* And until they are able to unravel the mysteries of this immaterial agent, their labors are fruitless, as regards the operations of *life.*

It is clear that all animals are sprung from others of the same species, and consequently modeled after the ones originally made.

Since the command was given to "increase and multiply," the propagation of animals, from the most insignificant animalcule to man, the highest in creation, will go on by fixed and immutable laws, without further interference on the part of the Creator. There will be no pause or intermission, until the edict for the suspension of further generation is proclaimed.

With a view of elucidating some of the laws which regulate

the workings of the organic nervous system, I will endeavor to point out the mode in which cutaneous perspiration is induced.

A familiar example is exhibited, when a person takes violent or active exercise. The respiration is rapidly increased; the heart pulsates violently; there is general arterial excitement, with a flushed countenance. The cause of all these changes consists in more oxygen passing into the blood, by the excitation of the pulmonary nerves, than is required for the maintenance of life in the organic nervous system. To preserve life, the excess of oxygen must be disposed of. The temperature of the blood is raised to a high degree; the serum of the blood gives off hydrogen, which combines with the surplus oxygen; serum is formed, which transudes through the pores of the skin, coming off in large drops. Hence the thirst, and demand for cold water to drink, is easily understood; hydrogen must be supplied for the oxygen.

It is now evident that the water which is imbibed by the mouth passes off by the skin. This proposition cannot be denied.

If water can pass out by the pores of the skin, there is no reason why the water should not pass in by the same inlets. It is on this principle that if a man, when thirsty, immerses himself in fresh or salt water, his thirst will subside. The hydrogen of the water unites with the excess of oxygen, the organic nerves are no longer over-stimulated, and the burning thirst ceases.

Hence it is, a man drenched with water will live for a long time, in comparison with another wearing *dry garments*, when both are similarly circumstanced—as the deprivation of water to drink. The danger of drinking ice-water, when the body is heated. is now susceptible of explanation. The cold paralyzes the organic nerves in the stomach—the transition from heat to cold produces a violent shock, which pervades the organic nervous system in an instant. Hence it is a person may drop dead.

To illustrate the manner in which the " wear and tear" of the organic nervous system is provided for, it is necessary to state, in the first instance, that fibrin is required for the reparation of a wound, that lime is requisite for the union of a fractured bone.

When a girl is suffering from chlorosis, the extreme paleness, the green tinge under the lower eyelids, the white scleroties, the dilated pupils, the blanched lips, the waxy and languid expression of countenance, the feeble gait, the palpitation of the heart, the pain in the side, the tendency to faint, the sense of suffocation on making exertion, the peculiar longings for certain loathsome kinds of food, the weakness of the back, cold extremities, the total or partial suppression of the menstrual flux, or the leucorrhœa, which very frequently exists—leave no doubt but the organic nervous system is deranged, and is in a state of inanition. The blood is in such cases impoverished, and destitute of iron, which, on being administered with other appropriate *adjuvantia*, quickly restores the patient to health and vigor. Here the iron produces its good effects, by repairing the material substance of the organic nervous system, exactly in the same way that *lime* or fibrin are conducive to the production of bone or development of muscle.

In the second stage of phthisis, when the tubercles in the lung or lungs are producing irritation or inflammation, profuse nocturnal perspirations harass the patient. He will be attacked with fever every evening, and complain in the morning of having wetted three or four shirts during the night. Under these circumstances, the respiration will be found hurried; the excitation of the pulmonary nerves will cause too much oxygen to pass into the blood, which must be liberated or set free by the process already specified, namely, perspiration.

In the last stage of phthisis, matters are quite different—when, forsooth, pneumo-thorax takes place, leaving only one lung to discharge the functions of respiration, and that too, perhaps, half destroyed by ulceration. It is here the accumulation of mucus rapidly fills the air-cells, and ushers in the mucous rattles in the bronchial tubes—keeping out the air, and consequently the oxygen, until at length the breath of life is extinguished.

If the spheno-palatine ganglion preside over the secretion of the saliva from the salivary glands, and the mucus from the tonsils, and that the secretion from the submaxillary gland take place through its connection with the Vidian nerve, I presume must be admitted.

Then follows, as a consequence, that the saliva of a rabid dog owes its poisonous influence to a morbid condition of the spheno-palatine ganglion; and that the difficulty, an insurmountable obstacle, presented to swallowing can be accounted for, by recollecting the parts, supplied with nerves from the spheno-palatine ganglion, are in a highly irritated and sensitive state.

If the spheno-palatine ganglion is the original seat of the disease, and the organ morbidly affected in the dog, it follows, as a legitimate consequence, that the spheno-palatine should be the ganglion implicated in a man laboring under hydrophobia; and that such is the case cannot be denied by any person, who has seen a case of hydrophobia. · *Verbum sat.*

It is almost unnecessary to remark that the spheno-palatine ganglion guides the poisonous secretion from all venomous beings.

It must be now obvious, a thorough and comprehensive knowledge of the laws and connections which govern and regulate the animal and organic nervous systems is indispensably required by every medical practitioner—such, in reality, being the alpha and omega of medical and surgical science. It is the foundation on which a permanent superstructure, capable of containing a universal knowledge of the nature of diseases, as well as a true explanation of the *modus· operandi* of therapeutic agents, can be erected.

It is, to use the words of a great philosopher and accomplished scholar, Professor Martyn Paine, "what will ultimately distinguish the scientific from the superficial physician."

N. B.—The foregoing chapters are printed from the AMERICAN MEDICAL GAZETTE, in which they were published, at various periods, without any alteration. Some sentences may appear objectionable, but as the papers were written under peculiar circumstances, to which it is unnecessary to advert, or explain, it has been deemed fit to let the articles stand as originally written.